TOXIC ORGANIC VAPORS

in the

WORKPLACE

Frank E. Jones

LEWIS PUBLISHERS
Boca Raton   Ann Arbor   London   Tokyo

Library of Congress Cataloging-in-Publication Data

Jones, Frank E.
      Toxic organic vapors in the workplace / Frank E. Jones.
          p. cm.
      Includes bibliographical references and index.
      ISBN 0-87371-900-X (alk. paper)
      1. Organic compounds—Environmental aspects. 2. Gases, Asphyxiating and poisonous—
Measurement. 3. Air—Pollution—Measurement. 4. Industrial hygiene. I. Title.
      TD885.5.074J66 1993
      628.5'3—dc20   93-1838
                                                                                                                        CIP

This book contains information obtained from authentic and highly regarded sources. Re-printed material is quoted with permission, and sources are indicated. A wide variety of references are listed. Reasonable efforts have been made to publish reliable data and information, but the author and the publisher cannot assume responsibility for the validity of all materials or for the consequences of their use.

© 1994 by CRC Press, Inc.
Lewis Publishers is an imprint of CRC Press

No claim to original U.S. Government works
International Standard Book Number 0-87371-900-X
Library of Congress Card Number 93-1838
Printed in the United States of America   2 3 4 5 6 7 8 9 0
Printed on acid-free paper

It is a privilege to dedicate this book to Horace A. Bowman, a truly great man who has profoundly influenced my entire professional career.

And the Lord God formed man of the dust of the ground
and breathed into his nostrils the breath of life; and man
became a living soul.

Genesis 2:7

Preface

There is, and will continue to be, much interest and activity in the identification and measurement of toxic organic gases and vapors in the workplace air. It is the purpose of this book to review the various methods, devices, and materials used to sample for, identify, determine, analyze for, and measure toxic organic vapors in the workplace. The emphasis is on recent literature; in-depth historical treatment of the various subjects is not attempted. Toxicology is outside the emphasis of the book and of the competence and primary interest of the author and is thus not treated in detail.

It is intended and hoped that the material presented here will be of interest and assistance primarily to industrial hygienists and to specialists in air pollution, indoor air quality, the indoor environment, occupational health, occupational safety, respiratory protection, industrial ventilation, laboratory ventilation, environmental pollution, environmental exposure, environmental health, environmental safety, environmental and occupational risk, environmental science, environmental engineering, environmental research, environmental testing, ecology, health science, and toxicology.

The book begins with a chapter on gas chromatography (GC) separately and combined with mass spectrometry (GC-MS) and further combined with computer (GC-MS-COMP). These techniques have been used quite extensively to identify and measure toxic organic gases and vapors. Therefore, many of the techniques and applications are covered. Where available, details are given of the parameters and their values in the various applications.

In the second chapter, Tenax, an adsorbent that is used extensively in sampling for toxic organic gases and vapors in air, is reviewed. Both Tenax GC, a polymer of 2,6-diphenylhydroquinone, made by oxidative coupling of 2,6-diphenyl phenol, and Tenax TA, a further development of Tenax GC, are treated.

In Chapter 3, sorbent tubes, tubes containing different sorbents, are reviewed. Sorbent tubes are used widely for collecting organics in the workplace air. The tubes and similar devices are used in a wide variety of structures and applications.

Since respirators are used by workers in many workplace environments for respiratory protection, the performance and use of respirator cartridges and canisters are reviewed in Chapter 4. A testing protocol for respirator cartridges and canisters, adsorptive capacity, effects of humidity and contaminant concentration, service life, and the applications of models, equations, and adsorption isotherms are reviewed.

Passive or diffusive sampling of air and its uses in personal monitoring of toxic gases and vapors are reviewed in Chapter 5. The treatment begins with a general discussion of diffusion. The use of the Fuller, Schettler, and Giddings relation to estimate diffusion coefficients for diffusion of vapors or gases in air is discussed with an example calculation for the diffusion coefficient for water vapor in air. A review follows of the experimental determination of the binary diffusion coefficients for 147 organic vapors and other vapors diffusing into air. The early development of a diffusion passive personal sampling device is reviewed, followed by various aspects and applications of diffusive sampling. Finally, two reviews of passive dosimetry and diffusive sampling are summarized.

At this writing, the subject of environmental tobacco smoke, and particularly the health consequences of exposure of nonsmokers to environmental tobacco smoke, is of intense interest. Since one of the most prevalent sources of air pollution (of particular concern indoors in the workplace, in the home, in restaurants, in public buildings, in buses, on trains, etc.) is smoke from cigarettes (by far the most prevalent of smoking pollutants are those from cigarette smoking), cigars, and pipes, Chapter 6 concentrates on environmental tobacco smoke and organic gases therefrom. Environmental tobacco smoke and its constituents, mainstream smoke and sidestream smoke, are defined and discussed. Among other topics discussed are human exposure to environmental tobacco smoke; proxies, surrogates, tracers, and markers; exposure to environmental tobacco smoke; benzene in smoking; and nicotine. Perhaps one of the most impressive and significant of findings is that "smoking is by far the largest anthropogenic source of background human exposure to benzene." *

Since gasoline is widely used as a motor fuel, there is interest in determining the exposure of consumers and workers in the gasoline industry to components of gasoline. Chapter 7 reviews this subject in detail. Following a review of a gasoline vapor sampling method evaluation, reviews are given of exposures to gasoline vapor of consumers when refilling their automotive gasoline tanks, exposures at refinery terminals, exposures at tanker/barge loading facilities, exposures at a service plaza, exposures at a high-volume service station, and exposures at petroleum bulk terminals. Monitoring, sampling, standards, and analysis are covered in these reviews.

Perhaps unusual in books related to industrial hygiene, Chapter 8 reviews methods for the detection and identification of chemical warfare agents, prima-

---

* Hattemer-Frey, H.A., C. C. Travis, and M. L. Land. "Benzene: Environmental Partitioning and Human Exposure," *Environ. Res.* 53:221–232 (1990).

rily in air. After a listing of such methods, experimental determinations are reviewed.

After an introduction citing the causes of polycyclic aromatic hydrocarbon (PAH) formation, the sources of PAHs, and the forms of degradation of PAHs, Chapter 9 reviews experimental determinations of PAHs. Sampling and analysis for experimental determinations are reviewed.

Formaldehyde is an important industrial chemical. In Chapter 10, the uses of formaldehyde, the need for its collection and analysis, exposure to formaldehyde, and the collection and measurement of formaldehyde in air are reviewed.

In developing this review, I am indebted to the many authors and publishers of the hundreds of publications that contribute to this work. In addition, I would like to convey my sincere appreciation to Christine Winter at CRC Press for her help and technical expertise in editing and reviewing and for rendering exemplary cooperation and encouragement throughout the entire production of this book.

Frank E. Jones
32 Orchard Way South
Potomac, MD 20854

The Author

 **Frank E. Jones** is currently an independent consultant. He received his bachelor's degree in physics from Waynesburg College, his master's degree in physics from the University of Maryland, and has also pursued doctoral studies in meteorology at the University of Maryland. He served as a physicist at the National Bureau of Standards (now the National Institute of Standards and Technology, NIST) in many areas including standardization for chemical warfare agents, chemical engineering, processing of nuclear materials, evaporation of water, humidity sensing, evapotranspiration, cloud physics, earthquake research, mass, length, time, flow measurement, volume, and sound.

Mr. Jones began work as an independent consultant upon his retirement from NIST in 1987. He is the author of more than 75 technical publications, two published books, and two other books that are now in preparation, and he also holds two patents. He is an Associate Editor of the National Council of Standards Laboratories Newsletter. He is a member of the American Industrial Hygiene Association, the Institute for Nuclear Materials Management, the Instrument Society of America, and ASTM, and is associated with other technical societies from time to time as they are relevant to his interests.

Table of Contents

TOXIC ORGANIC VAPORS

in the

WORKPLACE

CHAPTER 1

Gas Chromatography, Mass Spectrometry, Computer

INTRODUCTION

Gas chromatography-mass spectrometry (GC-MS) and GC-MS with computer (GC-MS-COMP) are used extensively to identify and measure toxic organic gases.

Compounds are separated by gas chromatography based on their volatility and interaction with a stationary phase.[1] The stationary phase ranges from molecular sieves to synthetic organic polymers and liquid crystals. The mobile phase is a gas.[2]

GC combines the inherent ability of the chromatographic column to separate individual components and the ease of interfacing to a variety of detection devices; it separates compounds — it does not provide detection.[2] A variety of detectors interfaced to the GC provide either nonspecific detection of compounds eluting from the column or highly specific and highly selective detection.[2]

In this chapter, the applications of GC, GC-MS, and GC-MS-COMP to the identification and measurement of toxic organic gases will be reviewed.

APPLICATIONS

Study of the Efficiencies of Thermal Desorption from Activated Carbon

A gas chromatograph was used to determine vapor concentrations in an experimental study of the efficiencies of thermal desorption of two-component organic solvents from activated carbon.[3]

In this study, 200 mg of activated carbon particles (12/30 mesh) were put into a 3-mL vial, and two kinds of solvents were injected directly using a microsyringe. After being left for 24 hours at room temperature, carbon particles were put into a heated column and the solvents were desorbed under a flow of nitrogen gas. The desorption column was made from a 4-mm I.D. glass column.

The column outlet gas was sampled at intervals of 1 to 2 min by a gas auto sampler, and the vapor concentrations were determined using a gas chromatograph fitted with a flame ionization detector (FID).

The operating parameters for the GC were

1. Column — 2 m × 3 mm glass column
2. Column packing — 15% 1,2,3-Tris(2-cyanoethoxy)propane or 15% polyethylene glycol 6000 on Uniport P
3. Column temperature — 70 to 120°C
4. Injection port temperature — 150°C
5. Detector — flame ionization detector (FID)

Thermal desorption time was 20 min. After thermal desorption, carbon particles were returned to the vial again. For desorption of residual solvents, 2 mL of carbon disulfide was used. Determination of desorbed solvents was by GC.

Heated column outlet gas containing desorbed vapor was sampled intermittently. "The shapes of desorption curves for the first solvent component in the two-component system were steeper than those for the pure component, but curves for the second component were similar to those for the pure component, except during the first few minutes. Desorption efficiencies for the first component increased as the initially adsorbed amount of the second component increased, but those from the second component were almost independent of the adsorbed amount of the first component."[3]

Laboratory Evaluation of Sorbent Tubes

A study was made by Bishop and Valis[4] to evaluate several multisorbent tubes designed for general air sampling for their analytical application toward thermal desorption into a gas chromatograph-mass selective detector system (GC-MSD).

Known amounts of a diverse mixture of organic vapors were adsorbed onto the adsorbent tubes and nitrogen atmospheres were passed through them to simulate air sampling, followed by thermal desorption into a GC-MSD system to determine the percent recoveries of the organics from the tubes. A GC was interfaced with a thermal desorption system and a mass selective detector.

The adsorbent tubes were made of Pyrex tubing 11.5 cm long × 6 mm O.D. × 4 mm I.D. The tubes contained a glass frit in the front end.

Tube A contained, in order, 90-mg sections of:

1. 60/80 mesh Tenax
2. 40/70 mesh Ambersorb XE-340
3. 60/80 mesh activated charcoal

Tube B contained:

1. A 300-mg front section of 20/40 mesh Carbotrap C
2. A 200-mg middle section of 20/40 mesh Carbotrap B
3. A final 125-mg section of 45/60 mesh Carbosieve II

Tube C contained:

1. A 125-mg front section containing 80/100 mesh Chromosorb 106
2. A 100-mg middle 20/40 mesh Carbotrap section
3. A final 75-mg 45/60 mesh Carbosieve SII section

A single preconcentration tube, used throughout the study, was similar to the sample tubes except that the I.D. was 2 mm rather than 4 mm. It was packed with, in order:

1. 20 mg of Tenax TA
2. 50 mg of Carbotrap
3. 50 mg of 80/100 mesh Porasil
4. 20 mg of activated charcoal

The parameters of the gas chromatograph were

1. Column — 30 m × 0.25 mm I.D. SPB-1 column
2. Column coating film thickness — 1 μm
3. Carrier gas — helium
4. Carrier gas flow — 1 mL/min
5. Oven temperature program — 35°C for 4 min, then to 100°C at 5 /min, then to 160°C at 20°C/min, at 160°C for 5 min
6. GC run time — 25 min

Integrated peak areas for the sample tubes were compared to peak areas obtained from standards.

The thermal desorption technique was found to be a viable procedure for qualitative and quantitative analysis with potential wide application. It was shown to be precise and reproducible, with capability for detection of 5 ng of a compound in air.

Determination of Respirator Cartridge Variability

A gas chromatograph fitted with a flame ionization detector has been used in a study of the determination of organic vapor respirator cartridge variability in terms of degree of activation of the carbon and cartridge packing density.[5]

By evaluating the shape of peaks resulting from injection of carbon tetra-chloride onto a gas chromatographic column packed with active carbon, adsorption isotherms were plotted. The adsorption isotherms were transformed, using the Dubinin/Radushkevich isotherm equations, to characteristic curves. This procedure was used to determine the degree of carbon activation.

The experimental adsorption isotherms were measured using an elution chromatographic technique.[6,7]

The operating parameters for the GC were

1. Column — 1/8 in. stainless steel column
2. Column packing — 0.1 g of 60/65 mesh carbon
3. Carrier gas — nitrogen
4. Carrier gas flow rate — 21–22 mL/min
5. Oven temperature — 200°C
6. Detector — flame ionization detector
7. Injection volume of carbon tetrachloride — 1 μL
8. Detector temperature — 200°C

The conclusions for the study included:

1. There was significant variability in the amount and degree of carbon activation between respirator cartridges of the three different manufacturers and among respirator cartridges of an individual manufacturer.
2. For the five different lots of cartridges examined, the packing density ranged from 0.40 g/cm^3 to 0.53 g/cm^3.

Determination of Trace Organics in Gas Samples Collected by Canister

An analytical method was developed for the determination of trace organics in samples collected by canister using a gas chromatograph-mass spectrometer (GC-MS).[8]

The canister used to collect samples was SUMMA-passivated to yield representative air samples. In the SUMMA passivation process, a pure chrome-nickel oxide layer is coated on the inner metal surface of the canister. The chrome-nickel oxide layer increases the stability and the storage interval of many organic compounds.

The operating parameters for the GC-MS were

1. Column — 6 m × 0.75 mm O.D. Supelco VOCOL megabore column
2. Carrier gas — helium
3. Column temperature program — 40°C for 4 min, then programmed to 180°C at 4°C/min
4. Scan range of GC-MS — m/z 35–300 at 1 sec/scan in electron impact mode

A cryogenic trap containing liquid argon, with a center tube of large internal diameter, was used to trap moisture from canister air.

It was concluded that the method works well for analysis for volatile organic compounds of ambient air samples collected in canisters.

Evaluation of Desorption Efficiency Determination Methods

A gas chromatograph was used in the evaluation of desorption (from activated charcoal) efficiency determination and in a comparison of three desorption sample preparation techniques, using acetone as a test compound.[9]

The three desorption sample preparation techniques were

1. Direct liquid spike
2. Direct vapor spike
3. Phase equilibrium

Standard NIOSH-approved charcoal tubes were used in the spiking of the test samples. The front section of the tubes contained 100 mg of charcoal, and the backup section contained 50 mg. The acetone loadings of the tubes ranged from 0.05 to 4.5 mg; an attempt was made to match the amount of acetone loaded on the charcoal for each of the sample preparation techniques.

The front sections (containing 100 mg of charcoal) of the liquid and vapor spike charcoal tubes were placed in an autosampler vial and 1.0 mL of carbon disulfide was added. After sealing, the vials were shaken for 30 min, in preparation for GC analysis. The backup sections were prepared similarly in order to detect any breakthrough. One sample, only, spiked at 4.0 mg of acetone, contained acetone at 1% of the level found in the front section.

The operating parameters of the gas chromatograph were

1. Column — 60 m × 0.75 mm I.D. Supelcowax 10 glass capillary
2. Column coating film thickness — 1 μm
3. Carrier gas — helium
4. Flow rate of carrier gas — 20 mL/min
5. Oven temperature — 50°C isothermal
6. Detector — flame ionization detector

7. Detector temperature — 300°C
8. Injector volume — 1 μL
9. Injector temperature — 250°C

From each preparation method, the experimental data were fitted to a linear model which was used to calculate the overall desorption efficiency. The three spiking methods gave results that were significantly different at the 5% error level. As the amount of acetone loading was increased, the differences between the methods became more apparent. The results for the vapor and liquid spikes were similar, the results for the phase equilibrium method did not appear to reflect desorption mechanism. The vapor method best simulated actual field samples.

Adsorptive Displacement Analysis on Activated Carbon

Improvements in sensitivity that were attained for an analysis method for multiple trace contaminants on activated carbons extracted by adsorptive displacement were described.[10] Adsorptive displacement is equilibration in a solvent, such as dichloromethane, containing a large excess of a strongly adsorbing solute/displacer such as benz[a]-anthracene-7,12-dione. In this paper, the lower limits for detection and analysis were extended downward to the low-ppm range, based on carbon. The improvement came from adjustments in the gas chromatographic technique, making it possible to analyze for components at 0.1 ng.

In this improved method, preloaded carbon samples were prepared by adding the carbon to a solution of multiple adsorbates in methylene chloride. After equilibration for at least 48 hours, the solvent was slowly evaporated to dryness and then the preloaded carbon was shaken with a solution of displacer. After equilibration for at least 7 days, drained supernatant liquid was cleared by centrifuging and a measured sample of the cleared solution was immediately spiked with a measured volume of acenaphthene internal standard solution and stored for analysis.

The spiked solution was analyzed in a gas chromatograph. The operating parameters for the GC were

1. Column — 60 m × 0.75 mm (wide bore) glass capillary
2. Column coating — methyl silicone-phenyl silicone
3. Detector — flame ionization detector

The chromatographic results were used to plot adsorption isotherms (mg adsorbate/g of carbon vs. equilibrium concentration in mg/L).

Detection of dibenz[a,h]anthracene at loadings below 10 μg/g of carbon was made, with no detectable irreversible adsorption.

Characterization of Carbon Molecular Sieves and Activated Carbon

A gas chromatograph was used in a comparative evaluation of various carbon adsorbents and their adsorbent properties and physical characteristics, in a defined working range.[11]

The nine carbon molecular sieves and the activated coconut charcoal used for the comparative evaluation were

1. Ambersorb XE-340
2. Ambersorb XE-347
3. Carboxen-563
4. Carboxen-564
5. Carboxen-569

6. Carbosieve S-III
7. Carbosieve S-I
8. Purasieve
9. Supercarb
10. Activated Charcoal

Four adsorbates, chosen to provide comparative information about the working ranges of the ten adsorbents were

1. Ethane
2. Vinyl chloride
3. Dichloromethane
4. Water

To simulate adsorbent tubes typically used in air monitoring, a 6 mm O.D. × 4 mm I.D. × 10 cm long adsorbent tube was connected to the injector and detector ports of a GC. Adsorbent bed weights of 0.2500 ± 0.0002 g were chosen to parallel typical adsorbent tube bed weights.

The operating parameters for the GC were

1. Carrier gas — helium
2. Flow rate of carrier gas — 30 mL/min
3. Oven temperatures — four different oven temperatures, chosen to provide retention times of 0.15 to 0.75 min for the four adsorbates
4. Detector — thermal conductivity detector

The authors concluded that the gas-solid chromatographic technique provided an effective comparative evaluation of the adsorption properties of the ten adsorbents; and allows, based on the molecular size of the airborne organic contaminants of interest, selection of an adsorbent tube to function in a known working range.

Activated Carbon Performance for Sequential Adsorbates

A gas chromatograph was used to determine the adsorption characteristics of a packed bed of carbon granules exposed to benzene and carbon tetrachloride sequentially, under dynamic flow conditions.[12]

The activated carbon used in the packed bed was a 6/10 mesh granular carbon. Kinetic adsorption tests were carried out on a vapor adsorption test apparatus that had been modified to permit the generation and subsequent mixing of up to three vapors simultaneously.

The operating parameters for the GC were

1. Column — 6 ft × 1/8 in. I.D. glass column
2. Column packing — 28% Pennwalt 223 + 4% KOH on 80/100 mesh Gas Chrom R support
3. Detector — flame ionization detector

Two sets of experiments were performed. In the first set of experiments, 2- to 3-g portions of characterized activated carbon were exposed to benzene or carbon tetrachloride for a time period equal to 11% of the time required for the vapor concentration downstream of the bed of activated carbon to reach 1% of the inlet vapor concentration. The activated carbon was then exposed to the other of the two vapors until the concentration of this vapor downstream of the activated carbon reached 1% of the upstream vapor concentration.

In the second set of experiments, the bed of activated carbon was exposed to the vapor of the first adsorbate for from 25 to 40% of the time required for 1% breakthrough. The activated carbon was then exposed to vapor of the second adsorbate until 1% breakthrough was achieved.

For the first set of experiments, the adsorption space occupied by the first adsorbate was estimated by rearranging the Wheeler adsorption equation.[13] For the second adsorbate, how much adsorption space that remained available for the second adsorbate was determined.

For the second set of experiments, the breakthrough time for a second adsorbate, after the activated carbon bed had been partially saturated with another vapor, was predicted.

It was found that the total adsorption space of the activated carbon was invariant at a fixed temperature and a fixed relative pressure. The authors concluded that this finding allowed prediction of activated carbon performance for adsorbates introduced concurrently and/or sequentially.

Thermal Desorption of Solvents Sampled by Activated Charcoal

A gas chromatograph was used for analysis of analytes desorbed by a thermal desorption apparatus used to desorb solvent vapors collected on activated charcoal.[14] The results of an optimization study of operating conditions and the desorption yields for 34 commonly used solvents were reported.

By introducing 500 mg of activated charcoal into glass tubing of 140 mm length, 8 mm O.D., and 6 mm I.D., sampling tubes were prepared. The charcoal layer was approximately 4 cm long. The tubes, before use, were purged at 350°C under nitrogen flow of 200 mL/min for 20 min. The tubes were then stored at ambient temperature in screw-capped vials.

The desorption apparatus consisted essentially of a stainless steel reservoir, a desorption oven, a vacuum pump, and associated hardware. A carrier gas containing the desorbed analytes passed into a GC through a gas sampling valve.

The operating parameters of the GC were

1. Column — 80 cm × 1/8 in. stainless steel column
2. Column packing — Carbopack C 0.1% SP 1000, 80/100 mesh
3. Detector — flame ionization detector

The solvent or solvent mixture under test was injected at a known rate into a stream of air to prepare standard atmospheres. The air was ambient air that had been drawn through an activated charcoal filter, humidified to approximately 70 or 80% relative humidity, passed through a heated glass tube (the evaporator), and finally was sent through the sampling tube held at room temperature (20 to 25°C).

The desorption yield from the apparatus was determined for 34 solvents distributed in 10 different mixtures. Seven sampling tubes were loaded for each of the mixtures investigated.

The desorption yields were generally higher than 90% and the reproducibility of the yields was generally better than 2%; they were largely independent of the mixture composition, the relative humidity of the sampled air, and the concentration, in the range usually observed in industrial hygiene practice.

Automatic Cryogenic Preconcentration/Cryofocusing System

An automated cryogenic preconcentration/cryofocusing system followed by a gas chromatography system has been used in the evaluation of sampling and analytical methods for monitoring toxic organics in air.[15]

The toxic organic samples were collected in passivated "SUMMA" stainless steel canisters. The samples were analyzed using a GC. The operating parameters for the GC were

1. Column — 50 m × 0.32 mm I.D., DB-1 fused-silica capillary column
2. Carrier gas — helium
3. Carrier gas flow rate — 1.2 cm^3/min
4. Makeup gas — nitrogen
5. Makeup gas flow rate — 26 cm^3/min
6. Column temperature program — 35°C for 4 min, then programmed at 6°C/min to 150°C
7. Detectors — flame ionization detector (FID) and electron capture detector (ECD) in parallel
8. FID gases
   a: hydrogen at a flow rate of 30 cm^3/min
   b: air at a flow rate of 400 cm^3/min
9. Sample flow rate — 20 cm^3/min for 10 min cooled at about −160°C

The system of analysis of air samples by cryogenic preconcentration followed with a selective detection (FID/ECD) was found to be adequate if the compounds of interest were known. If not, GC-mass spectrometry was recommended for initial screening.

Effect of Adsorbed Water on Solvent Desorption from Activated Carbon

The effect of adsorbed water on solvent desorption of organic vapors collected on activated carbons was studied by Rudling and Bjorkholm[16] with analysis using a gas chromatograph with a flame ionization detector.

Two charcoals (activated carbons) were used in the experimental work. The BET surface areas of the activated carbons were 1120 and 1150 m^2/g, the total pore volumes were 0.59 and 0.60 cm^3/g, and the micropore volumes were about 0.43 and 0.37 cm^3/g. Both activated carbons were dried before use.

Samples were prepared by charging 100 mg of the charcoal into 3-mL screw cap vials. Water was added to the samples, and about 1 hour later, analyte was added. The analyte was added by allowing the liquid to vaporize from the tip of a microliter syringe. With the vial placed in a horizontal position, larger amounts of analyte were injected on the walls of the vial; the analyte entered the carbon as a vapor, using this procedure.

Desorption of the samples was with 1.0 mL of desorbing solvent followed by shaking for a minimum of 45 min.

No details of the gas chromatographic procedures were given by the authors; they said that standard techniques were used.

Water adsorption isotherms were determined experimentally for the two charcoals.

Analytes used were

1. *n*-Butanol
2. *p*-Dioxane
3. Ethyl Cellosolve (2-ethoxyethanol)

Carbon disulfide was used as the desorbing agent in every experiment. The authors concluded that:

1. The desorption efficiency for water-soluble compounds can be changed by adsorbed water, if the desorbing agent used is carbon disulfide
2. The above effect depends on the amount adsorbed and the water/carbon disulfide ratio
3. The carbon surface also is affected by the presence of water; the problem can be eliminated by the use of a desorbing agent capable of dissolving water

Determination of Styrene Exposure

Gas chromatographs have been used in the study of a commercially available diffusive tube sampler, for determination of exposures to styrene.[17]

The sampler used consisted of a stainless steel tube, 90 mm long × 6 mm O.D. × 5 mm I.D. The tube was packed with 200 mg of Tenax GC. The unpacked tube is commercially available, the user is left with the choice and packing of the adsorbent.

A syringe injection technique[18] was used to generate standard atmospheres of styrene vapor. The standard atmospheres were checked according to the United Kingdom Health and Safety Executive (HSE) diffusive sampler protocol.[19] The diffusive tube samplers were exposed in replicates of six The uptake rates were calculated from the mass of styrene adsorbed, the delivered concentration, and the exposure time. The range of concentrations and exposure times were typical of those used in the determination of occupational exposure.

The charcoal tubes were desorbed using 1 mL of carbon disulfide containing an internal standard, ethyl benzene. The desorbed solutions were analyzed using either of two gas chromatographs. The operating parameters provided by the authors were

1. Column — 20 m SE-54 capillary column
2. Mode — split

For thermal desorption, the desorption parameters were

1. Desorption — 2 stage
2. Desorption temperature — 250°C
3. Desorption time — 10 min
4. Column — 20 m SE-54 capillary column
5. Carrier gas — helium
6. Carrier gas flow rate — 1 cm³/min
7. Split — 50:1
8. Column temperature — 100°C
9. Styrene retention time — 3 min

Following the protocol[19] again, extensive field trials were made. The field trials were made in a variety of occupational hygiene situations in which workers might be potentially exposed to styrene vapor.

There were two types of field trials. In the first type of field trial, one of the diffusive tube samplers and one of an independent charcoal tube were exposed simultaneously, usually on the lapel of a person at work. This type of field trial was designed to determine field bias.

In the second set of experiments, replicates of each type of sampler were tested, usually in a static position. This type of field trial was designed to determine the precision of each sampling method.

The authors concluded that:

1. The diffusion sampler tube containing Tenax GC could be used to determine styrene in air.
2. A coefficient of variation of about 12% was found.
3. A coefficient of variation for replicate field samples was about 6%.
4. There was no appreciable loss of styrene from the diffusion sampler tubes during storage at room temperature for up to 6 months.
5. The diffusion sampler tubes were not affected by humidity.
6. The diffusion sampler tubes could be used as static samplers as well as personal samplers.

Gas Chromatographic Determination of Atmospheric Peroxyacetyl Nitrate (PAN)

Two gas chromatographs with electron capture detectors (GC-ECD) and one gas chromatograph using a luminol-chemiluminescence based detector device were used in a 24-day intercomparison study of peroxyacetyl nitrate (PAN) measurements.[20]

A chemiluminescence NO_x-based calibration technique and a NO_2-calibrated luminol-PAN GC were evaluated in the intercomparison study in a rural site in central Ontario, Canada between February 28 and March 23 of 1989. PAN samples, the concentrations of which were determined ultimately by infrared spectroscopy, were used to calibrate one of the PAN GC-ECDs; a dynamic mode chemiluminescence technique was used to calibrate the PAN GC-ECD. Standard PAN samples ranging in level between about 0.2 to 15 ppb, prepared by two independent methods, were exchanged among the three GCs. During the intercomparison study, ambient PAN concentrations ranged between 0.1 and 2.0 ppb.

The operating parameters for the GC-ECD referred to as the York University GC-ECD were

1. Column —105 cm × 6.3 mm O.D. glass column
2. Column packing — 10% Carbowax 400 on Chromosorb G-AW
3. Carrier gas — 5% methane/95% argon
4. Carrier gas flow rate — 65 mL/min
5. Column temperature — 35°C
6. Detector — electron capture detector
7. Detector temperature — 50°C
8. PAN retention time — about 3.9 min

The operating parameters for the GC-ECD referred to as the Atmospheric Environment Services GC-ECD were

1. Column — 60 cm × 6.3 mm O.D. glass column
2. Column packing — 5% Carbowax 400 on Chromosorb G-AW
3. Carrier gas — Ultra high purity nitrogen
4. Carrier gas flow rate — 60 mL/min
5. Column temperature — 40°C
6. Detector — electron capture detector
7. Detector temperature — 60°C
8. PAN retention time — about 5.3 min

The operating parameters for the GC referred to as the Unisearch Associates GC included:

1. Column — 35 cm × 6.3 mm O.D. Teflon column
2. Column packing — 10% Carbowax 400 on Chromosorb G-AW
3. Carrier gas — partially purified ambient air
4. Carrier gas flow rate — 160 cm³/min
5. Column temperature — 37°C
6. Detection system — luminol-based chemiluminescence detection system
7. PAN retention time — about 2 min

The results of analysis of standard samples were within ±25% of the standard concentration, for all 3 GCs. The authors considered that the results of the intercomparison study (1) "demonstrated the reliability of the chemilumines-cence calibration technique," and (2) "indicated that the luminol-based PAN measurement device can yield accurate and sensitive measurements of ambient PAN concentrations."[20]

Solid Sorbent for Crotonaldehyde in Air

A gas chromatograph was used in a study of 13X sieves as a solid sampling medium for crotonaldehyde in air.[21]

Molecular sieves, 13X of two different mesh sizes, 8–12 mesh and 60–80 mesh, were chosen for the study. The 8–12 mesh sieves were activated by heating slowly for 21 hours at 400°C with a flow of nitrogen of 25 mL/min through the sieves. The 60–80 mesh sieves were activated by the manufacturer. Glass sampling tubes were packed with weighed amounts of the sieves, held in place by glass wool plugs.

By the addition of appropriate amounts of glass-distilled crotonaldehyde to distilled water, standards were prepared.

A gas chromatograph was used to analyze gas mixtures, liquid desorbates, and liquid standards. The operating parameters of the chromatograph were

1. Column — Carbopack B with a 5% Carbowax 20M liquid phase
2. Carrier gas — nitrogen
3. Detector — flame ionization detector

Among the conclusions drawn from the study were

1. 13X Molecular sieves were excellent sorbents for crotonaldehyde vapors.
2. Sampling of air with a sieve tube was simple, practical, and effective.
3. Desorption of the molecular sieves with distilled water gave quantitative recovery.
4. Collected samples were stable for at least 1 week when stored at low temperature.
5. Sample volume was restricted by high relative humidities.

Worker Exposure to Polycyclic Aromatic Hydrocarbons

A gas chromatograph connected to a mass spectrograph was used in laboratory development of sampling and analytical methods for determining the physical nature of worker exposure to polycyclic aromatic hydrocarbons.[22] The method consisted of filter and sorbent tube sampling followed by benzene extraction.

The operating parameters for the GC-MS were

1. Column — 30 m × 0.25 mm DB-5 column
2. Column coating film thickness — 0.25 μm
3. Injection mode — splitless
4. Carrier gas — helium
5. Carrier gas flow rate — 1 mL/min
6. Oven temperature program — 60°C for 1 min, then programmed at 10°C/min to 280°C, then at 280°C for 20 min
7. Injector temperature — 280°C
8. GC-MS detection mode — multiple ion detection mode

The method was used to "demonstrate that workers in paving and roofing operations and on some worksites in the steel and silicon carbide industries show an exposure profile that suggests minimal health risk and is largely different from the exposure of workers in aluminum refineries, refractory brick laying and most other worksites in the silicon carbide industry."[22]

Use of High-Resolution Gas Chromatography-Mass Spectroscopy in a Comparison in the Analysis of Norwegian Soil for Nine Polycyclic Aromatic Hydrocarbons

To study the distribution of nine polycyclic aromatic hydrocarbons (PAHs) in Norwegian soil, high-resolution gas chromatography-mass spectroscopy (HRGC-MS) was used for analysis.[23]

Preliminarily, to enable comparison of the pattern of 9 PAHs from air samples to the pattern for soil samples, 8 randomly selected 24-hour air samples were collected using a high-volume sampler.[24] Soil samples were collected from 12 locations.

The PAHs selected for analysis were

1. Naphthalene
2. Acenaphthene
3. Biphenyl
4. Fluorene
5. Phenanthrene
6. Fluoranthene
7. Pyrene
8. Chrysene/triphenylene
9. Benzo[a]pyrene

Internal standards used were deuterated biphenyl and 9-methyl-anthracene.

The operating parameters of the quadrapole HRGC-MS used for the qualitative (multiple ion mode) and quantitative (selected ion mode) analysis of soil samples were

1. Column — 25 m × 0.25 mm I.D. DB-5 Chrompack column
2. Average phase thickness — 0.25 μm
3. Oven temperature program — 50°C for 0.2 min, raised to 100°C at 25° C/min, then raised to 295°C (to remain for 5 min) at 8°C/min
4. Injector temperature — 295°C
5. Inlet flow rate — 66.3 mL/min

The operating parameters of the HRGC used for analysis of spiked test samples were

1. Column — 25 m × 0.25 mm I.D. DB-5 Chrompack column
2. Average phase thickness — 0.25 μm
3. Oven temperature program — 50°C for 0.2 min, raised to 100°C at 25°C/min, then raised to 295°C (to remain for 5 min) at 8°C/min

The HRGC-MS analysis showed that, for individual unsubstituted PAHs, the concentrations in soil ranged from less than 1 ppb (the detection limit) to 993 ppb.

Gas Chromatography for Laboratory Analysis and Field Validation of a Method for Monitoring Short-Term Exposures to Ethylene Oxide

Gas chromatography was used for analysis in the laboratory and field evaluation of a JXC charcoal sampling and analytical method for monitoring short-term exposures to ethylene oxide.[25]

In the laboratory, charcoal in JXC charcoal tubes was desorbed, using carbon disulfide, and allowed to stand 30 min with stirring. Analysis of the eluent was made with a gas chromatograph with a flame ionization detector (GC-FID).

In the field validation, two types of samples were collected from a field atmosphere with JXC charcoal tubes. The results were compared with those obtained from a portable gas chromatograph. There was excellent agreement between the GC-FID data and charcoal mean data. The pooled bias, assuming the GC-FID data to be the known concentration, was 3.9%. The calculated coefficient of variation for the laboratory validation was 5.6%; the pooled coefficient of variation for the field validation data was 8.1%.

Capillary Gas Chromatography in a Procedure for Determination and Quantitation of Ethylene Oxide, 2-Chloroethanol, and Ethylene Glycol

In a procedure for detection and quantitation of ethylene oxide, 2-chloroethanol, and ethylene glycol, capillary gas chromatography was used.[26] To the knowledge of the authors, this was the first capillary gas chromatographic method for the determination of the three above compounds with a single injection. Two capillary columns were used, one with an I.D. of 0.25 mm and one with an I.D. of 0.32 mm. Two different solvents were used: dimethylformamide and water.

The operating parameters for the GC with the narrower column (0.25 mm I.D.) with water as the solvent were

1. Column — 30 m × 0.25 mm I.D. DB-Wax (cross-linked and bonded polyethylene glycol-fused silica capillary) column
2. Carrier gas — helium
3. Carrier gas flow velocity — 34 cm/sec
4. Solvent — water
5. Injection mode — split, 1:100
6. Injection volume — 5 μL
7. Temperature program — 60°C for 2 min, then to 210°C at 16°C/min
8. Detector — dual flame detectors
9. Detector temperature — 300°C

The operating parameters for the GC with the narrower column with dimethylformamide as the solvent were

1. Column— 30 m × 0.25 mm I.D. DB-Wax column
2. Carrier gas — helium
3. Carrier gas flow velocity — 34 cm/sec
4. Solvent — dimethylformamide
5. Injection mode — split, 1:50
6. Injection volume —2 μL
7. Temperature program — 60°C for 2 min, then to 180°C at 16°C/min
8. Detector — dual flame detectors
9. Detector temperature — 300°C

The operating parameters for the GC with the wider column (0.32 mm I.D.) with on-column mode were

1. Column — 30 m × 0.32 mm I.D. DB-Wax column
2. Carrier gas — helium

3. Carrier gas flow velocity — 29 cm/sec
4. Solvent — water
5. Injection mode — on-column
6. Injection volume — 0.5 μL
7. Temperature program — 70°C for 1.5 min, then to 180°C at 12°C/min
8. Detector — dual flame detectors
9. Detector temperature — 300°C

The operating parameters for the GC with the wider column with split injection mode were

1. Column — 30 m × 0.32 mm I.D. DB-Wax column
2. Carrier gas — helium
3. Carrier gas flow velocity — 29 cm/sec
4. Solvent — water
5. Injection mode — split, 1:50
6. Injection volume — 5 μL
7. Temperature program — 60°C for 1.6 min, then to 175°C at 12°C/min
8. Detector — dual flame detectors
9. Detector temperature — 300°C

Quantitation of the three compounds, ethylene oxide, 2-chloroethanol, and ethylene glycol, could be obtained with a single GC analysis. Linearity of the three compounds was demonstrated.

Gas Chromatographic Determination of "Mustard Gas" or "Sulfur Mustard"

A gas chromatograph equipped with a flame photometric detector (GC-FPD) was used in gas chromatographic determination of bis(2-chloroethyl)sulfide, also known as "mustard gas" or "sulfur mustard", in air at trace concentrations.[27]

The mustard vapor was trapped on a bed of Tenax GC, transferred to a smaller sorbent bed, and thermally desorbed into a GC-FPD. Two distinct sets of conditions for the GC-FPD were used: one based on a packed chromatographic column and one based on a fused silica capillary column.

The operating parameters for the packed column were

1. Column — 6 ft long × 3 mm O.D. × 1.5 mm I.D. Teflon column
2. Column packing — 10% (weight/weight) OV-202 on 80/100 mesh Gas-Chrom Q
3. Carrier gas — nitrogen
4. Carrier gas flow rate — 20 mL/min
5. Flow rates for other gases — air:132 mL/min; oxygen:18 mL/min; hydrogen:102 mL/min
6. Oven temperature — 120°C
7. Detector — flame photometric detector in the sulfur-specific mode
8. Detector temperature — 200°C
9. Injection port temperature — 200°C

The operating parameters for the capillary column were

1. Column — 15 m × 0.53 mm I.D. DB-210 fused silica capillary column
2. Coating of stationary phase — 1.0 µm
3. Carrier gas — helium
4. Carrier gas flow rate — 20 mL/min
5. Flow rates for other gases — air:111 mL/min; hydrogen: 55 mL/min
6. Oven temperature — 85°C
7. Detector — flame photometric detector in the sulfur-specific mode
8. Detector temperature — 200°C
9. Injection port temperature — 200°C

The authors concluded that mustard vapor could be determined with high sensitivity, and with adequate accuracy and precision, by sampling through triethanolamine-impregnated filters onto Tenax GC sorbent beds; followed by thermal desorption of the analyte from the sorbent beds into a GC with a flame photometric detector.

Gas Chromatographic Determination of Ethyl 2-Cyanoacrylate

A gas chromatograph equipped with a thermionic specific detector (TSD) and a gas chromatograph-mass spectrometer (GC-MS) were used in the determination of ethyl 2-cyanoacrylate in the workplace environment.[28]

Air was drawn through a Teflon filter, then through a sorbent tube containing two sections of Tenax. Acetone was used to extract separately the filter and the Tenax from both sections of the sorbent tube. The extracts of ethyl 2-cyanoacrylate and methyl 2-cyanoacrylate were quantified by gas chromatography.

The operating parameters for the GCs were

1. Column — 1.8 m × 6 mm O.D. × 2 mm I.D. glass column
2. Column packing — 50/80 mesh Tenax GC
3. Carrier gas — nitrogen
4. Flow rate of carrier gas — 30 mL/min
5. Detector — thermionic specific detector in the nitrogen mode
6. Oven temperature — 200°C
7. Detector temperature — 230°C
8. Injector temperature — 220°C

The authors concluded that airborne ethyl 2-cyanoacrylate and methyl 2-cyanoacrylate could be monitored by drawing air through tubes containing Tenax followed by desorption with acetone and quantification for cyanoacrylate by gas chromatography with a thermionic specific detector or by selected ion monitoring using chemical ionization mass spectrometry.

Gas Chromatographic Analysis for Measuring *m*-Xylene

In a comparison of three sampling and analytical methods for measuring *m*-xylene in the expired air of exposed humans, a gas chromatograph was used for analysis.[29]

The breath of human subjects was sampled for *m*-xylene following controlled exposures to 75 ppm *m*-xylene in air for 4 hours. Breath was sampled using a stainless steel sampler that permitted continuous mainstream or sidestream sampling of solvents present: from the mainstream with a charcoal cloth sorbent and from the sidestream using Tenax TA.

The Tenax sampling tubes were packed with 100 mg of the sorbent in the front section and 50 mg in backup tubes. The operating parameters for the GC into which Tenax samples were thermally desorbed for analysis were

1. Column — 3 m × 2 mm I.D. stainless steel column
2. Column packing — 10% SP-2100 on 100/120 mesh Supelcoport
3. Carrier gas — nitrogen
4. Flow rate of carrier gas — 20 cm³/min
5. Column temperature — 93°C
6. Detector — flame ionization detector
7. Detector temperature — 255°C
8. Injector temperature — 205°C

Peak heights were used for the analysis of the Tenax samples.

The operating parameters for the GC used for the analysis of the charcoal cloth samples were

1. Column — 2.7 m × 2 mm I.D. stainless steel column
2. Column packing — SP-2100 on 100/120 mesh Supelcoport
3. Carrier gas — nitrogen
4. Flow rate of carrier gas — 30 cm³/min
5. Column temperature — 110°C
6. Detector — flame ionization detector
7. Detector temperature — 200°C
8. Injector temperature — 200°C

Peak areas were used for the analyses.

The operating parameters for the GC used for the analysis of alveolar samples were

1. Column — 2 m × 2 mm I.D. stainless steel column
2. Column packing — 3% SP-2100 on 100/120 mesh Supelcoport
3. Column temperature — 110°C
4. Detector — flame ionization detector
5. Detector temperature — 200°C
6. Injector temperature — 200°C

Either peak heights or peak areas were used to determine sample concentrations.

The precision for each sampling and analytical method was 0.13 for alveolar sampling, 0.14 for mainstream-mixed sampling (12 subjects), and 0.23 for sidestream-mixed sampling (12 subjects).

Gas Chromatography-Mass Spectrometric Analysis for Nitrosamines

To confirm the presence of *N*-nitrosamines in workplace air sample extracts, three capillary gas chromatography-mass spectrometry (GC-MS) procedures were developed.[30]

The first procedure utilized high resolution selected ion monitoring of NO^+ coupled with capillary GC for confirmation.

The second procedure incorporated capillary GC along with high resolution multiple ion detection for identification of specific *N*-nitrosamines.

The third procedure, with the mass spectrometer operated in the full scan mode, provided a complete mass spectrum of the compounds under study.

The 8 compounds studied were

1. *N*-nitrosodimethylamine
2. *N*-nitroso-*N*-methylethylamine
3. *N*-nitrosodiethylamine
4. *N*-nitrosodipropylamine
5. *N*-nitrosodibutylamine
6. *N*-nitrosopyrrolidine
7. *N*-nitrosopiperidine
8. *N*-nitrosomorpholine

A known volume of air was drawn through tubes to collect field samples on ThermoSorb/N sorbent tubes. Samples were backflushed with 1.8 to 2.0 mL of a 3:1 mixture of dichloromethane and methanol. The eluent was collected in vials; 5 µL of the eluent was analyzed by GC-thermal energy analysis; 1-µL aliquots were used for GC-MS analyses.

The operating parameters for the GC-MS system were

1. Column — 25 m × 0.21 mm I.D. fused silica capillary column
2. Column coating — 0.25-µm-thick film of Carbowax 20M
3. Carrier gas — helium
4. Flow rate of carrier gas — 46 cm/sec
5. Column temperature program — 55°C to 200°C, at 15°C/min
6. Mode — splitless

The detection limits, on a per injection basis, for the three GC-MS procedures were

1. Full scan GC-MS — 2 ng
2. Screening procedure — 10 to 20 pg
3. GC-MS High-Resolution Multiple-Ion Detection — 2 to 4 pg

The authors concluded that the limits of detection of the procedures were sufficient for typical workplace environmental samples, and that data obtained from the analysis of several samples indicated that these GC-MS procedures provided a highly reliable approach for confirmation of *N*-nitrosamines in samples of workplace air.

Dinitrotoluene on Activated Carbon

GC-MS has been used in a study of the adsorption of 2,4-dinitrotoluene (DNT) in aqueous solution by two commercial activated carbons and desorption of these adsorbents by solvent extraction.[31]

The characterization of chemicals desorbed from the spent carbons was made in the following sequence:

1. After desorption, the solution was filtered
2. The solvent was removed
3. The residue was dried under nitrogen for several hours
4. The dried residue was silylated
5. The silylated residue was analyzed using GC-MS

The operating parameters of the GC-MS were

1. Column — 30 m Durabond I narrow-bond fused silica column
2. Column temperature program — 90°C for 1 min, raised to 300°C at 10°C/min

The authors concluded that:

1. Both of the activated carbons proved to be excellent adsorbents for the removal of DNT from aqueous solution
2. Acetone and methanol were both effective solvents for extraction of DNT from spent carbons
3. Acetone was the most effective for the removal of chemicals produced during the adsorption other than DNT

In the extracted solutions, at least six chemicals other than DNT were present. Four of the identified chemicals were 2,4-dinitrobenzyl alcohol, 2,4-dinitrobenzaldehyde, 2,4-dinitrobenzoic acid, and 2,4-dinitrobenzoate.

Detection of Acetaldehyde and Acetone

A gas chromatograph was used in the testing of the response of a mercuric oxide Reduction Gas Detector (RGD-2) to subpicomole and larger quantities of acetaldehyde and acetone.[32]

A gas chromatograph was set up in such an arrangement that a flame ionization detector (FID) and an RGD-2 could be connected alternately to the exit of a packed column. The operating parameters of the GC were

1. Column — 2 m × 2.2 mm I.D. stainless steel column
2. Column packing — 0.2% Carbowax 1500 on Carbopack C
3. Column temperature — 60°C
4. Carrier gas — Ultrapure helium
5. Flow rate of carrier gas — 25 mL/min

Independent streams of makeup gas continuously flowed to each detector, the flow rate of the makeup gas was 15 mL/min. The detector that was not currently involved in the testing was continuously flushed with the makeup gas to prevent contamination from room air.

The sensitivities of the detectors were compared over a range of approximately 0.1 to 200 pmol for acetaldehyde and acetone. The sensitivity of the reduction gas detector was an improvement by approximately one order of magnitude over the sensitivity of the flame ionization detector. The reduction gas detector was easily adapted to a conventional gas chromatograph.

Analysis of Effluents from Heated Polyurethane Foam

A complicated mixture of decomposition products from rigid polyurethane foam heated to 80°C was analyzed by capillary gas chromatography with flame ionization detection, thermionic specific detection, electron capture detection, and mass selective detection.[33]

Rigid polyurethane foam was heated at 80°C in an aluminum tube. Zero air and high purity nitrogen were used separately as carrier gases for the effluent from the foam. The carrier gas then passed through an adsorber tube held at a temperature slightly above 0°C. The adsorber tube was a 75 mm × 6.3 mm O.D. tube containing a 2-cm column of Tenax TA of 35/60 mesh.

Effluent samples were analyzed on two gas chromatographs (GC). The operating parameters for the first GC were

1. Column — 5.30 m × 0.32 mm I.D. SPB column
2. Column film thickness — 0.25 μm
3. Carrier gas — helium
4. Flow rate of carrier gas — 25 cm/sec
5. Oven temperature program — –50°C, held for 10 min; –50°C to 30°C at 20°C/min, held for 5 min; 30°C to 250°C at 10°C/min, held for 1 min
6. Detectors
   a. Flame ionization detector-FID
   b. Electron capture detector-ECD
   c. Thermionic selective detector-TSD

The operating parameters for the second GC were

1. Column — 30 m × 0.32 mm I.D. DB-5 fused silica column
2. Column film thickness — 1 μm
3. Carrier gas — helium
4. Flow rate of carrier gas — 23 cm^3/sec
5. Oven temperature program — 30°C, held for 5 min; 30°C to 150°C at 10°C/min, held for 2 min
6. Detector — mass selective desorber (MSD)

More than 20 compounds were tentatively identified as off-gases. Off-gas emission rates were significantly higher in zero air atmospheres than in a nitrogen atmosphere.

Gas Chromatograph in Evaluation of Diffusional Dosimeter for Monitoring Methyl Chloride

A gas chromatograph was used in the evaluation of a thermally desorbable diffusion dosimeter for monitoring methyl chloride in the workplace.[34]

The diffusional monitor consisted of a cylindrical capsule containing 700 mg of the solid adsorbent Anasorb GM, held in place in a plastic badge. In the laboratory portion of the study, the passive monitors were exposed to methyl chloride in an exposure chamber. In the field portion of the study, monitoring was done side by side, using both the passive monitor and a validated charcoal tube method.[35] The purpose of the field study was to determine whether the two methyl chloride monitoring methods provided equivalent results.

The operating parameters for the gas chromatograph were

1. Column — 6 m × 3.2 mm O.D., 15% FFAP on 80/100 mesh, Gas Chrom Q column
2. Detector — hydrogen flame ionization detector

The diffusional monitor and the validated charcoal tube method produced statistically indistinguishable results when measurements were made side by side under field conditions. The author concluded that the diffusional monitor in combination with Anasorb GM solid adsorbent was a satisfactory means of determining methyl chloride exposures in the workplace.

Gas Chromatograph Used in the Evaluation of a Novel Gas and Particle Sampler

A gas chromatograph was used in the evaluation of a novel gas and particle sampler.[36] The sampler used a prototype denuder inlet adapted to a dichotomous sampler for the collection of organic compounds in the gas and particulate phases.

The diffusion denuder sampling system was tested using hexachlorobenzene and/or lindane in air. The operating parameters for the gas chromatograph were

1. Column — 2 m × 4 mm glass column
2. Column packing — 80/100 mesh high-performance Chromosorb W, coated with 3% OV-1 and 60% OV-215
3. Carrier gas — 5% methane in argon gas
4. Flow rate of carrier gas — 30 mL/min
5. Detector — electron capture detector
6. Column temperature — 190°C
7. Detector temperature — 350°C
8. Injector temperature — 250°C

The trapping efficiency for the target compounds, hexachlorobenzene and lindane, was found to be better than 98% over a 24-hour sampling period at representative ambient concentrations under most simulated atmospheric conditions.

Capillary Analyses of Volatile Pollutants

A gas chromatograph was used in a study of factors influencing analyses of volatile pollutants by wide-bore capillary chromatography and purge-and-trap techniques.[37]

The following factors were investigated:

1. Initial column temperature
2. Carrier gas flow rate
3. Speed with which pollutants are desorbed from the trap
4. Type of detector used

The operating parameters for the GC were

1. Column — 60 m × 0.75 mm I.D. wide-bore 1.5 μm VOCOL capillary column
2. Carrier gas — helium
3. Column temperatures
   a. For analyses of a mixture of 5 volatile gases: 10°C or 35°C
   b. For analyses of 35 volatile pollutants: held at 10°C for 6 min, then programmed to 170°C at 6°C/min
4. Carrier gas flow rates — for 3a above: 5, 10, and 15 mL/min; for 3b above: 10 mL/min

The authors concluded that once the four factors above were optimized, volatile pollutants were readily analyzed. A carrier gas flow of 10 mL/min and an initial column temperature of 10°C were the optimum conditions to use for purge-and-trap analyses of volatile priority pollutants from a 0.75-mm I.D. VOCOL capillary column. Chromatography of the volatile pollutants can be improved by the use of an experimental trap. For chromatographic analysis, the choice of detectors is also critical. Poor chromatography can be caused by large dead volume.

Modification of a Portable Gas Chromatograph

A portable gas chromatograph was modified to avoid the disadvantages of purge-and-trap systems and to improve chromatographic separation of volatile compounds.[38] The modification was the coupling of a capillary column with the portable GC system. The column was a 30 m × 0.75 mm I.D. borosilicate glass column, coated with an SPB-1 film 1 µm thick.

The detection limit for the modified system was determined to be 20 pg benzene. Identified peaks were confirmed by gas chromatography-mass spectrometry, demonstrating that the portable system could be used for both screening and qualitative or quantitative analysis with high precision.

The author concluded that the results demonstrated that the modified system could be used in the field at ambient temperature to obtain chromatograms with sufficient resolution for reliable sensitivities and quantitation in the ppb range.

Study of Decreased Sensitivity of Total Organic Vapor Analyzers

A portable gas chromatograph was used in a study of the reduction in sensitivity to toluene and gasoline in the presence of methane (0.5 to 5.0% v/v) of total organic vapor analyzers equipped with photoionization detectors (PID).[39]

The ability of three total organic vapor analyzers to detect toluene and gasoline standards in mixtures of methane and air was tested. A portable gas chromatograph and two total organic vapor detectors also were used to determine the extent of a leak from an underground gasoline tank. A screening column was used for all the field portable GC analyses.

For the laboratory study of the effects of methane, the operating parameters for the GC were

1. Column — 0.61 m SE30 column
2. Column temperature — ambient temperature
3. Carrier gas — hydrocarbon-free air

The results showed that in the presence of methane there was an exponential decrease in photoionization detector sensitivity. There was a reduction in sensitivity of about 30% for 0.5% methane and reduction in sensitivity of about 90% for 5% methane. The authors suggested that a flame ionization detector (FID) total organic vapor analyzer should be used to screen for methane or that a chromatographic column should be used to separate the compounds before users should rely on photoionization data.

Comparison of Portable Gas Chromatographs and Passivated Canisters

A comparison was made of portable gas chromatograph (PGC) data with passivated canister data for volatile organic compounds.[40]

Air samples and PGC calibration standards were collected in spherical 6-L electropolished canisters. A canister was held with its inlet less than 10 cm from the end of the PGC probe and the valve was opened to fill the probe during the time the PGC sample pump was running, in order to make direct comparison between canister and PGC results. A GC equipped with flame ionization and electron capture detectors was used to analyze air samples collected in canisters, transported to a laboratory, and cryogenically preconcentrated.

A microprocessor-controlled PGC was equipped with a constant-temperature column enclosure. The enclosure contained a 10 m × 0.53 mm I.D. wall-coated open-tubular column; 1 m of the column; was a back-flushable precolumn. A chemically bonded stationary liquid phase was used. The carrier gas was ultrazero air (containing less than 0.1 ppm carbon).

At U.S. and overseas field studies at industrial, hazardous waste, and roadway sites, data were obtained. The results for the field studies suggested to the authors that a combination of canister and PGC methods offer a synergistic approach to source assessment measurements.

The authors concluded that portable gas chromatographs can rapidly produce valid estimates of ambient background concentrations of many volatile nonpolar and semipolar organic air pollutants.

Evaluation of Occupational Exposure in Silicon Carbide Plants

A gas chromatograph-mass spectrometer (GC-MS) was used to measure polycyclic aromatic hydrocarbons (PAH) in field samples collected in the silicon carbide industry.[41]

The samples were collected using a sampling train that combined a glass fiber filter and a solid Chromosorb 102 adsorbent tube to trap both particulate and gaseous PAHs. Benzene or a 1% solution of methanol in carbon disulfide was used to extract particulates on the filter. The quantification of the PAHs was done by GC-MS.

The 12 PAHs quantified were

1. Naphthalene
2. Biphenyl
3. Acenaphthene
4. Fluorene
5. Phenanthrene
6. Anthracene
7. Fluoranthene
8. Pyrene
9. Chrysene
10. Benzo[a]anthracene
11. Benzo[e]pyrene
12. Benzo[a]pyrene

PAHs were found to be produced during the silicon carbide synthesis, but they were strongly adsorbed on graphite. The adsorption of PAHs on graphite prevented their determination as matter soluble in benzene.

Comparison of Air Monitoring Strategies

In a comparison of three air monitoring strategies for assessing personal exposure to volatile organic compounds, samples collected on Tenax GC sorbent cartridges and in SUMMA-polished canisters were analyzed by gas chromatography-mass spectrometry.[42]

Samples were collected at residential indoor, residential outdoor, and centralized locations. Samples collected on the Tenax GC cartridges were analyzed by thermal desorption-injection capillary GC-MS. Volatile organics were desorbed from the sorbent cartridges at 260°C with a nominal helium flow. Samples from canisters were analyzed by cryogenic trapping of approximately 100-mL aliquots followed by GC-MS-COMP analysis.

The authors concluded that "a centrally located sampling site cannot be used to predict outdoor residential (i.e., backyard) levels, which in turn cannot be used to predict indoor levels at the proximate residence."

Sampling and Analysis for Trace Volatile Organic Emissions from Consumer Products

Gas chromatographs were used in comparisons of two techniques for trace analysis of volatile organic compound emissions from consumer products.[43] The two techniques were (1) direct on-line sampling and analysis, and (2) on-line sorbent collection followed by off-line analysis.

Two types of direct analysis were examined: (1) direct injection of emissions from an environmental chamber loaded with sample into a gas chromatograph equipped with a flame ionization detector (GC-FID) for compound identification, and (2) direct injection of headspace-collected emissions into a gas chromatograph equipped with a mass selective detector.

Collection of the volatile organic emissions on a solid sorbent, followed by thermal desorption-gas chromatographic-mass spectrometric (TD-GC-MS) analysis was compared to both direct on-line methods.

The operating parameters for the GC-FID were

1. Column — 25 m × 0.32 mm I.D. SE 54 Ultra column
2. Column coating thickness — 0.5 μm
3. Carrier gas — nitrogen
4. Carrier gas linear velocity — 36 cm/sec
5. Column temperature program — –80°C isothermal for 2 min; to 150°C at 15°C/min, then isothermal for 10.0 min
6. Mode — splitless
7. Detector — flame ionization detector
8. Injector temperature — 150°C
9. Detector temperature — 300°C

The operating parameters for the GC-MS used for headspace analysis were

1. Column — a 25 m × 0.32 mm I.D. SE 54 Ultra column
2. Column coating thickness — 0.5 μm
3. Carrier gas — helium
4. Carrier gas linear velocity — 32 cm/sec
5. Column temperature program — –80°C isothermal for 2 min; to 300°C at 8°C/min, then isothermal for 15 min

6. Mode — splitless
7. Detector — mass selective detector (MSD)
8. Injector temperature — 150°C
9. MSD transfer line temperature — 286°C

The operating parameters for the GC-MS used for determination of organics collected on adsorbent tubes were

1. Column — 25 m × 0.32 mm I.D. SE 54 Ultra column
2. Carrier gas — helium
3. Carrier gas linear velocity — 57 cm/sec
4. Column temperature program — 30°C isothermal for 1 min; to 300°C at 8°C/min, then isothermal for 10 min

The authors concluded that (1) direct interfacing of an environmental chamber with a GC system apppeared to be an attractive method for consumer product emissions, and (2) TD-GC-MS was particularly useful for out-gassed products in extremely trace quantities and in a mid-boiling point range.

Permeation of Protective Clothing Materials

A gas chromatograph was used in a study of the permeation of methylene chloride and perchloroethylene through seven protective clothing materials.[44]
The operating parameters for the GC were

1. Column — Apezion column
2. Carrier gas — helium
3. Carrier gas flow rate — 30 mL/min
4. Column temperature — 150°C
5. Sensitivity of the GC — 0.3 µg/mL

For the protective clothing materials the following were determined:

1. Breakthrough time
2. Steady state permeation rate
3. Solubility in methylene chloride and perchloroethylene
4. Diffusion coefficients

Potential Exposure of Cooks

In a study of potential exposure of cooks to airborne mutagens and carcinogens, a gas chromatograph-mass spectrometer system was used.[45] The GC-MS was used to analyze eight air samples taken in four restaurants for carcinogens.

The air samples were taken with 37-mm Grade AA glass fiber filters and 600-mg Tenax sorbent tubes. An ultrasound bath, using cyclohexane and methanol in a 90:10 mixture, was used to extract filters and sorbents together for each sample.

The operating parameters for the GC-MS were

1. Column — 60 m S-WAX-10 capillary column
2. Oven temperature program — 50°C for 10 min; to 100°C at 40°C/min, then held for 2 min; to 270°C at 6°C/min
3. Carrier gas — helium
4. Mode — split or splitless depending on the concentration of the mixture
5. GC-MS interface temperature — 250°C

The GC-MS analyses identified a broad range of aliphatic and aromatic hydrocarbons with a variety of functional groups, many including oxygen and nitrogen, and one including chlorine. None of the substances identified was a known carcinogen.

Tetrachloroethene from Coin-Operated Dry Cleaning Establishments

A gas chromatograph was used in a study of tetrachloroethene air pollution originating from coin-operated dry cleaning establishments in the Federal Republic of Germany.[46]

Activated carbon tubes were used to collect air samples from (1) 15 coin-operated dry cleaning establishments (CODC), (2) one building where a CODC had been run, and (3) a private car transporting a dry-cleaned down jacket.

The activated carbon was eluted with toluene and the eluate was subsequently analyzed by gas chromatography. The operating parameters of the GC used were

1. Column — 50 m × 0.25 mm I.D. SE 54 column
2. Carrier gas — helium
3. Auxiliary gas — argon/methane
4. Column temperature — 80°C, isothermal
5. Mode — split-splitless
6. Detector — ^{63}Ni-electron capture detector
7. Detector temperature — 290°C
8. Injector temperature — 275°C

The tetrachloroethene (TCE) concentration was found to be

1. Between 3.1 and 331 mg/m^3 within the CODC
2. Slowly decreasing in the building in which a CODC had once been run (after removal of dry cleaning machines)
3. Up to 24.8 mg/m^3 in the car transporting the garment

Organic Pollutant Emissions from Unvented Kerosene Space Heaters

In an exploratory study of semivolatile and nonvolatile organic pollutant emission rates from unvented kerosene space heaters, gas chromatography and mass spectrometry were used for analysis of samples.[47]

Organic pollutants were collected on Teflon-impregnated glass filters backed by XAD-2 resin. A gas chromatograph with a flame ionization detector was used for total chromatographic analysis. A combined gas chromatograph-mass spectrometer was used for analysis for nitrated polycyclic aromatic hydrocarbons and other organics. An electron impact (EI) GC-MS provided tentative determinations of the various classes of compounds.

The results of the study showed that kerosene heaters can emit

1. Polycyclic aromatic hydrocarbons (PAHs)
2. Nitrated PAHs
3. Alkylbenzenes
4. Phthalates
5. Hydronaphthalenes
6. Aliphatic hydrocarbons
7. Alcohols
8. Ketones
9. Other organic compounds, some of which are known mutagens

Emissions from Unvented Coal Combustion

Gas chromatographs were used for a study of emissions from coal combustion in China.[48] A filter and an XAD-2 resin were used to collect particles and semivolatile organics.

Organic material was extracted from the filters and the XAD-2 resin by Soxhlet extraction with dichloromethane. Gas chromatography was used to analyze for total chromatographable organics. A GC-MS was used for analysis for determination of polycyclic aromatic hydrocarbons.

It was found that coal combustion emitted both direct- and indirect-acting mutagens, most of which were frameshift mutagens.

Analysis of Urban-Related Aquatic and Airborne Volatile Organic Compounds

Gas chromatography and gas chromatography-mass spectrometry were used in a study of the distribution of volatile organic compounds in urban-influenced air and river waters in Spain.[49]

Air samples were taken by parallel adsorption on charcoal and polyurethane foam. At a flow rate of 4 L/min, volumes of 0.1 to 1 m^3 were collected. Samples of subsurface river and marine water were collected by immersion of 2-L amber-glass bottles.

The operating parameters of the gas chromatograph on which all samples were analyzed were

1. Column — 45 m × 0.32 mm I.D. SE-52 fused-silica capillary column
2. Column coating thickness — 0.45 μm
3. Carrier gas — hydrogen
4. Carrier gas linear velocity — 0.5 m/sec
5. Mode — splitless
6. Column temperature program — 30°C for 1 min; to 50°C at 40°C/min; to 220°C at 3°C/min
7. Detector — flame ionization detector
8. Detector temperature — 250°C
9. Injector temperature — 225°C

Selected samples were analyzed by GC-MS.

The authors concluded that parallel air sampling with charcoal and polyurethane foam is needed to cover a range of volatile organic compounds similar to that afforded by the closed-loop stripping technique in water.

Determination of Isocyanate and Aromatic Amine Emissions

Gas chromatography was used in studies of the use of 4,4′-methylenediphenyl isocyanate (MDI)-based polyurethane-bound foundry core materials.[50]

For gas chromatographic analysis, air samples were drawn through an impinger bottle containing 10 mL of 0.4 M hydrochloric acid. The acid solution was alkalized with saturated sodium hydroxide, and amines were extracted resulting in an amide solution. The amide solution was analyzed by GC with the use of electron capture detection, nitrogen selective detection, or mass spectroscopic detection. The analytical results showed one value for each isocyanate and its corresponding amine.

Laboratory studies and field investigations were made for two foundries in Sweden. The field investigations demonstrated the presence of all of the anilines found in the laboratory studies, although the amount of methylenedianiline was below the detection limit of 0.001 mg/m³.

The authors concluded that phenyl isocyanates, and especially anilines, were potential high-level pollutants in foundries where MDI-based polyurethane binders were used, and that both isocyanate and amine selective methods should be used when monitoring pollutants in atmospheres of foundries in which polyurethane binders were used.

Concentrations of Volatile Organic Compounds at a Building

A gas chromatograph-mass spectrometer system was used in the determination of concentrations of volatile organic compounds in a facility with a history of occupant complaints; symptoms were characteristic of "sick building syndrome".[51]

Samples of air were collected in the four buildings in the facility using a 3M organic vapor monitor or passive sampler. Each badge was spiked with an internal standard prior to analysis. The badge was then extracted using 1.5 L of carbon disulfide. The organic compounds in the extract were separated and identified using GC-MS procedures.

The operating parameters for the GC-MS used were

1. Column — 12 m × 0.2 mm narrow-bore, corss-linked dimethyl silicone, fused-silica capillary column
2. Carrier gas — helium
3. Carrier gas flow rate — 1 mL/min
4. Mode — split
5. Column temperature program — 0°C to 220°C at 8°C/min
6. Injection port temperature — 240°C

More than 40 different organic compounds with concentrations in excess of 1 µg/m^3 were identified; for several species, the concentrations were greater than 100 µg/m^3. Cleaning products, floor wax, latex paints, and reentrained motor vehicle exhaust were sources of the identified compounds. The dominant source was the hydraulic system for the elevators of the buildings; some compounds being volatilized from the hydraulic fluid used in the system.

Polynuclear Aromatic Compounds in Environmental Samples Collected at a Coal Gasification Process Development Unit

GC-MS was used in evaluation of a synchronous luminescence screening procedure performed at the Oak Ridge National Laboratory to rank coal gas gasification process samples collected at a coal gasification process development unit (PDU).[52]

Three series of samples were taken: (1) 10 personal air samples collected at a coal gasification PDU, (2) 18 methylene chloride extracts of XAD-2 resin used for collection of some organic compounds in the product gas, and waste-water samples from a coal gasification PDU, and (3) 11 XAD-2 resin samples, water, and solid samples from a coal gasification PDU.

There was good agreement between the results obtained by the synchronous luminescence technique and by detailed GC-MS.

Identification of Bisphenol-A by GC-MS

Bisphenol-A in the thermal degradation products of epoxy powder paint was identified by GC-MS.[53]

The operating parameters for the GC-MS used were

1. Column — 25 m × 0.2 mm I.D. BP-5 column
2. Carrier gas — helium

3. Carrier gas flow rate — 0.8 mL/min
4. Column temperature program — 35°C for 1 min; to 240°C at 6°C/min; at 240°C for 20 min
5. Injected volume — 1 μL

The maximum formation of bisphenol-A was about 0.02% of the sample weight. The generation of bisphenol-A was found to be temperature dependent.

Performance Evaluation of a Gasoline Vapor Sampling Method

Gas chromatography was used in a study conducted to evaluate the performance of a method used by Exxon to monitor exposures of workers to gasoline vapors.[54]

An aliquot of the headspace above liquid gasoline that was at equilibrium at 20°C in a closed chamber was charged into a 950-L stainless steel test chamber, at 20°C and less than 40% relative humidity. The chamber atmosphere was drawn through a sorbent train consisting of two charcoal tubes in series: a large charcoal tube, 400/200, followed by a small charcoal tube, 100/50.

A gas chromatograph was used to analyze the charcoal tubes. The operating parameters of the GC were

1. Column — 30 ft × 1/8 in. stainless steel packed column
2. Column temperature program — 35°C, then increased at 5°C/min
3. Detector — flame ionization detector

The results of the evaluation indicated that the large, 600-mg charcoal tube yielded excellent results if the sample flow rate was adjusted properly with regard to absolute humidity.

REFERENCES

1.   Jennings, W. *Analytical Gas Chromatography* (Orlando, FL: Academic Press, 1987).
2.   *Exposure Assessment for Airborne Pollutants: Advances and Opportunities* (Washington, D.C.: National Research Council, National Academy of Sciences, 1991).
3.   Hori, H., I. Tanaka, and T. Akiyama. "Thermal Desorption Efficiencies of Two-Component Organic Solvents from Activated Carbon," *Am. Ind. Hyg. Assoc. J.* 50:24–29 (1989).
4.   Bishop, R. W., and R. J. Valis. "A Laboratory Evaluation of Sorbent Tubes for use with a Thermal Desorption Gas Chromatography-Mass Selective Detection Technique," *J. Chromatogr. Sci.* 28:589–593 (1990).
5.   Trout, D., P. N. Breysee, T. Hall, M. Corn, and T. Risby. "Determination of Organic Vapor Respirator Cartridge Variability in Terms of Degree of Activation of the Carbon and Cartridge Packing Density," *Am. Ind. Hyg. Assoc. J.* 46:491–496 (1986).

6. Dollimore, D., G. R. Heal, and D. R. Martin. "An Improvement in Elution Technique for Measurement of Adsorption Isotherms by Gas Chromatography," *Chromatography* 50:209–218 (1970).

7. Habgood, H. W. "Chromatography and the Solid-Gas Interface," in *The Solid Gas Interface*, Vol. 2, Ed. E. A. Flood (New York: Marcel Dekker, 1967), pp. 611–646.

8. Hsu, J. P., G. Miller, and V. Moran, III. "Analytical Method for Determination of Trace Organics in Gas Samples Collected by Canister," *J. Chromatogr. Sci.* 29:83–88 (1991).

9. McCurry, J. D., I. Stoll, K. M. Mitchell, and R. D. Zwickl. "Evaluation of Desorption Efficiency Determination Methods for Acetone," *Am. Ind. Hyg. Assoc. J.* 50:520–525 (1989).

10. Thakkar, S., and M. Manes. "Adsorptive Displacement Analysis of Many-Component Priority Pollutants on Activated Carbon. 2. Extension to Low Parts per Million (Based on Carbon)," *Environ. Sci. Technol.* 22:470–472 (1988).

11. Betz, W. R., S. G. Maroldo, G. D. Wachob, and M. C. Firth. "Characterization of Carbon Molecular Sieves and Activated Charcoal for Use in Airborne Contaminant Sampling," *Am. Ind. Hyg. Assoc. J.* 50:181–187 (1989).

12. Jonas, L. A., and E. B. Sansone. "Prediction of Activated Carbon Performance for Sequential Adsorbates," *Am. Ind. Hyg. Assoc. J.* 47:509–511 (1986).

13. Wheeler, A., and A. J. Robell. "Performance of Fixed-Bed Catalytical Reactors with Poison in Feed," *J. Catal.* 24:13:299–305 (1969).

14. Cocheo, V., G. G. Bombi, and R. Silvestri. "An Apparatus for the Thermal Desorption of Solvents Sampled by Activated Charcoal," *Am. Ind. Hyg. Assoc. J.* 48:189–197 (1987).

15. Jayanty, R. K. M. "Evaluation of Sampling and Analytical Methods for Monitoring Toxic Organics in Air," *Atmospheric Environment* 23:777–782 (1989).

16. Rudling, J., and E. Bjorkholm. "Effect of Adsorbed Water on Solvent Desorption of Organic Vapors Collected on Activated Carbon," *Am. Ind. Hyg. Assoc. J.* 47:615–620 (1986).

17. Brown, R. H., K. J. Saunders, and K. T. Walkin. "A Personal Sampling Method for the Determination of Styrene Exposure," *Am. Ind. Hyg. Assoc. J.* 48:760–765 (1987).

18. Health and Safety Executive. *Methods for the Determination of Hazardous Substances. Generation of Test Atmospheres by the Syringe Injection Technique (MDHS 3)* (London: Health and Safety Executive, 1981).

19. Brown, R. H., R. P. Harvey, C. J. Purnell, and K. J. Saunders. "A Diffusive Sampler Evaluation Protocol," *Am. Ind. Hyg. Assoc. J.* 45:67–75 (1984).

20. Blanchard, P., P. B. Shepson, K. W. So, H. I. Schiff, J. W. Bottenheim, A. J. Gallant, J. W. Drummond, and P. Wong. "A Comparison of Calibration and Measurement Techniques for Gas Chromatographic Determination of Atmospheric Peroxyacetyl Nitrate (PAN)," *Atmos. Environ.* 24A:2839–2846 (1990).

21. Mann, J. H., Jr., and A. Gold. "A Solid Sorbent for Crotonaldehyde in Air," *Am. Ind. Hyg. Assoc. J.* 47:832–834 (1986).

22. Lesage, J., G. Perrault, and P. Durand. "Evaluation of Worker Exposure to Polycyclic Aromatic Hydrocarbons," *Am. Ind. Hyg. Assoc. J.* 48:753–759 (1987).

23. Vogt, N. B., F. Brakstad, K. Thrane, S. Nordenson, J. Krane, E. Aamot, K. Kolset, K. Esbensen, and E. Steinnes. "Polycyclic Aromatic Hydrocarbons in Soil and Air: Statistical Analysis and Classification by the SIMCA Method," *Environ. Sci. Technol.* 21:35–44 (1987).

24. Thrane, K. E., and A. Mikalsen. "High-Volume Sampling of Airborne Polycyclic Aromatic Hydrocarbons Using Glass Fibre Filters and Polyurethane Foam," *Atmos. Environ.* 25:909–918 (1991).

25. Puskar, M. A., J. L. Nowak, and L. H. Hecker. "Laboratory and Field Validation of a JXC Charcoal Sampling and Analytical Method for Monitoring Short-Term Exposures to Ethylene Oxide," *Am. Ind. Hyg. Assoc. J.* 49:237–243 (1988).

26. Danielson, J. W., R. P. Snell, and G. S. Oxborrow. "Detection and Quantitation of Ethylene Oxide, 2-Chloroethanol, and Ethylene Glycol with Capillary Gas Chromatography," *J. Chromatogr. Sci.* 28:97–101 (1990).

27. Fowler, W. K., and J. E. Smith, Jr. "Solid Sorbent Collection and Gas Chromatographic Determination of Bis(2-chloroethyl)sulfide in air at Trace Concentrations," *J. Chromatogr. Sci.* 28:118–122 (1990).

28. Gaind, V. S., and K. Jedrzejczak. "Gas Chromatographic Determination of Ethyl 2-Cyanoacrylate in the Workplace Environment," *Analyst* 114:567–569 (1989).

29. Glaser, A. A., J. E. Arnold, and S. A. Shulman. "Comparison of Three Sampling and Analytical Methods for Measuring *m*-xylene in Expired Air of Exposed Humans," *Am. Ind. Hyg. Assoc. J.* 51:139–150 (1990).

30. Cooper, C. V. "Gas Chromatographic/Mass Spectrometric Analysis of Extracts of Workplace Air Samples for Nitrosamines," *Am. Ind. Hyg. Assoc. J.* 48:265–270 (1987).

31. Ho, P. O., and C. S. Daw. "Adsorption and Desorption of Dinitrotoluene on Activated Carbon," *Environ. Sci. Technol.* 22:919–924 (1988).

32. O'Hara, D., and H. B. Singh. "Sensitive Gas Chromatographic Detection of Acetaldehyde and Acetone Using a Reduction Gas Detector," *Atmos. Environ.* 22:2613–2615 (1988).

33. Krzymien, M. E. "Analysis of the Effluents from Polyurethane Foam Heated at 80°C," *Am. Ind. Hyg. Assoc. J.* 48:67–72 (1987).

34. Hahne, R. M. A. "Evaluation of the GMD Systems, Inc., Thermally-Desorbable Diffusional Dosimeter for Monitoring Methyl Chloride," *Am. Ind. Hyg. Assoc. J.* 51:96–101 (1990).

35. Hahne, R. M. A. "Validation of a Thermally Desorbable Passive Monitor for Methyl Chloride," (Intern. rep.) Midland, MI: The Dow Chemical Company, 1989.

36. Lane, D. A., N. D. Johnson, S. C. Barton, G. H. S. Thomas, and W. H. Schroeder. "Development and Evaluation of a Novel Gas and Particle Sampler for Semivolatile Chlorinated Organic Compounds in Ambient Air," *Environ. Sci. Technol.* 22:941–947 (1988).

37. Mosesman, N. H., L. M. Sidisky, and S. D. Corman. "Factors Influencing Capillary Analyses of Volatile Pollutants," *J. Chromatogr. Sci.* 25:351–355 (1987).

38. Jerpe, J. "Ambient Capillary Chromatography of Volatile Organics With a Portable Gas Chromatograph," *J. Chromatogr. Sci.* 25:154–157 (1987).

39. Nyquist, J. E., D. L. Wilson, L. A. Norman, and R. B. Gammage. "Decreased Sensitivity of Photoionization Detector Total Organic Vapor Detectors in the Presence of Methane," *Am. Ind. Hyg. Assoc. J.* 51:326–330 (1990).

40. Berkley, R. E., J. L. Varns, and J. Plell. "Comparison of Portable Gas Chromatographs and Passivated Canisters for Field Sampling Airborne Toxic Organic Vapors in the U.S. and the U.S.S.R.," *Environ. Sci. Technol.* 25:1439–1444 (1991).

41. Dufresne, A., J. Lesage, and G. Perrault. "Evaluation of Occupational Exposure to Mixed Dusts and Polycyclic Aromatic Hydrocarbons in Silicon Carbide Plants," *Am. Ind. Hyg. Assoc. J.* 48:160–166 (1987).

42. Michael, L. C., E. D. Pellizzari, R. L. Perritt, and T. D. Hartwell. "Comparison of Indoor, Backyard, and Centralized Air Monitoring Strategies for Assessing Personal Exposure to Volatile Organic Compounds," *Environ. Sci. Technol.* 24:996–1003 (1990).

43. Bayer, C. W., M. S. Black, and L. M. Galloway. "Sampling and Analysis Techniques for Trace Volatile Organic Emissions from Consumer Products," *J. Chromatogr. Sci.* 26:168–173 (1988).

44. Vahdat, N. "Permeation of Protective Clothing Materials by Methylene Chloride and Perchloroethylene," *Am. Ind. Hyg. Assoc. J.* 48:646–651 (1987).

45. Teschke, K., C. Hertzman, C. Van Netten, E. Lee, B. Morrison, A. Cornista, G. Lau, and A. Hundal. "Potential Exposure of Cooks to Airborne Mutagens and Carcinogens," *Environ. Res.* 50:296–308 (1989).

46. Gulyas, H., and L. Hemmerling. "Tetrachloroethene Air Pollution Originating from Coin-Operated Dry Cleaning Establishments," *Environ. Res.* 53:90–99 (1990).

47. Traynor, G. W., M. G. Apte, H. A. Sokol, J. C. Chuang, W. G. Tucker, and J. L. Mumford. "Selected Organic Pollutant Emissions from Unvented Kerosene Space Heaters," *Environ. Sci. Technol.* 24:1265–1270 (1990).

48. Mumford, J. L., D. B. Harris, K. Williams, J. C. Chuang, and M. Cooke. "Indoor Air Sampling and Mutagenicity Studies of Emissions from Unvented Coal Combustion," *Environ. Sci. Technol.* 21:308–311 (1987).

49. Rosell, A., J. I. Gomez-Belinchon, and J. O. Grimalt. "Gas Chromatographic-Mass Spectrometric Analysis of Urban-Related Aquatic and Airborne Volatile Organic Compounds. Study of the Extracts Obtained by Water Closed-Loop Stripping and Air Adsorption with Charcoal and Polyurethane Foam," *J. Chromatogr.* 562:493–506 (1991).

50. Renman, L., C. Sango, and G. Skarping. "Determination of Isocyanate and Aromatic Amine Emissions from Thermally Degraded Polyurethanes in Foundries," *Am. Ind. Hyg. Assoc. J.* 47:621-628 (1986).

51. Weschler, C. J., H. C. Shields, and D. Rainer. "Concentrations of Volatile Organic Compounds at a Building with Health and Comfort Complaints," *Am. Ind. Hyg. Assoc. J.* 51:261–268 (1990).

52. Abbott, D. W., R. L. Moody, R. M. Mann, and T. Vo-Dinh. "Synchronous Luminescence Screening for Polynuclear Aromatic Compounds in Environmental Samples Collected at a Coal Gasification Process Development Unit," *Am. Ind. Hyg. Assoc. J.* 47:379–385 (1986).

53. Peltonen, K., P. Pfaffli, A. Itkonen, and P. Kalliokoski. "Determination of the Presence of Bisphenol-A and the Absence of Diglycidyl Ether of Bisphenol-A in the Thermal Degradation Products of Epoxy Powder Paint," *Am. Ind. Hyg. Assoc. J.* 47:399–403 (1986).

54. Russo, P. J., G. R. Florky, and D. E. Agopsowicz. "Performance Evaluation of a Gasoline Vapor Sampling Method," *Am. Ind. Hyg. Assoc. J.* 48:528–531 (1987).

CHAPTER 2

Tenax

INTRODUCTION

Tenax GC is an aromatic polyether, a polymer of 2,6-diphenylhydroquinone, made by oxidative coupling of 2,6-diphenyl phenol.[1] It has also been expressed as a polymer based on 2,6-diphenyl-para-phenylene oxide[2] or 2,6-diphenyl-*p*-phenylene oxide.[3]

Among the advantages of Tenax GC are

1. It has a high affinity for organic compounds[4]
2. It is regarded as the best porous polymer adsorbent when relatively high-boiling components are of interest[5,6]
3. It has a very high temperature stability of 380 to 400°C[7,8]
4. It is fairly hydrophobic, and high recoveries of volatiles are obtained quickly on thermal elution[4,9]
5. Its thermal stability accounts for relatively low background levels[10]
6. It is claimed to be inert with respect to sampled compounds[11]
7. No chemical reactions were observed in the presence of substantial amounts of ozone[12]

However, the presence of certain artifacts has been established.[10] Even after careful cleanup of the polymer, some artifacts remain.[13]

In an attempt to overcome the artifact problem, Tenax TA became available.[13] Tenax TA is a further development of Tenax GC.[14] It can be cleaned simply by heating in an inert gas flow.[15,16]

Tenax adsorbent has a widespread use in the field of air analysis.[17,18] It is widely used for collection of volatile organic compounds, VOCs, in ambient air.[19,20]

Due to the widespread use of Tenax GC and Tenax TA in the areas of interest in this book, this chapter is devoted to the relevant uses of Tenax.

SAMPLING OF ORGANIC COMPOUNDS IN THE PRESENCE OF INORGANIC GASES

The Tenax GC sampling cartridge was tested to gain insight into potential problems which may be encountered during field sampling.[18]

The Tenax GC sampling cartridge was a glass cartridge with a bed length of 6.0 cm and an I.D. of 1.5 cm, through which air was drawn by a sampler. The cartridges were spiked with authentic standards using permeation tubes or a solvent evaporation method.

A thermal desorption gas chromatography system with a flame ionization detector was used for qualitative analysis of decomposition products.

Experiments were run on the following:

1. Effect of O_3, NO_x, and humidity on Tenax GC
2. Effect of O_3, NO_x, and SO_2 on Tenax GC
3. Effect of molecular chlorine
4. Reaction of adsorbed analytes with oxidants
5. Inhibition of *in situ* reactions on Tenax GC

The results from the experiments were

1. There is a region in which a combination of high concentration of oxidants and high relative humidity appeared to reduce the amounts of the three decomposition products (benzaldehyde, acetophenone, and phenol), by decomposing them faster than they were generated from the adsorbent
2. When sampling ambient air with a high relative humidity, benzaldehyde, acetophenone, and phenol were more prone to increase in magnitude
3. The total concentration of oxidants in air reaching the Tenax GC sorbent was important in producing the decomposition products
4. An increase in the decomposition products or compounds specific to the presence of sulfur oxides could be detected under the conditions of sulfur dioxide exposure studied
5. There was an increase in acetophenone relative to the other decomposition products when Tenax GC cartridges were exposed to ozone, chlorine, and humidity

The authors concluded that the results of the experiments clearly demonstrated that the presence of oxidants in ambient air can produce decomposition products both from the Tenax GC sorbent and from analytes adsorbed to the sorbent.

A COMPARISON OF ARTIFACT BACKGROUND ON THERMAL DESORPTION OF TENAX GC AND TENAX TA

A comparison was made of the artifact background on thermal desorption of Tenax GC and Tenax TA.[21]

The Tenax was packed into silanized acid (concentrated hydrochloric acid) washed glass tubes. The Tenax GC used in the study was preconditioned under a nitrogen flow of 50 mL/min at 275°C for 24 hours. The Tenax TA to be packed in a first set of tubes was preconditioned as above. The Tenax TA to be packed in a second set of tubes was preconditioned under nitrogen at 340°C for 2 hours. Capillary gas chromatography was used to analyze single and pooled tubes packed with 200 mg of 60/80 mesh Tenax.

Over a period of 1 min, the Tenax was heated to 250°C and then maintained at 250°C for 1 min. Helium carrier gas transferred any heat-desorbed volatiles into a liquid nitrogen cold trap and the helium flow was maintained for 10 min after heating. Then the cold trap was removed from the liquid nitrogen and fed into the oven of the gas chromatograph. As far as possible, the components of the heat-desorbed volatiles were identified by gas chromatography-mass spectrometry.

The results of the study showed that the most efficient preconditioning procedure for Tenax, in order to minimize artifact background on thermal desorption, was heating of the adsorbent under purified nitrogen at 340°C for 4 hours and then heating at 300°C for two separate periods of 15 min immediately before use. The blank chromatograms for Tenax TA were superior to those for Tenax GC.

Of the volatiles identified by combined gas chromatography-mass spectrometry analysis of pooled blank tubes, the majority were aliphatic, alicyclic, and aromatic hydrocarbons. Many of the volatiles were components of common food aromas. The pooling technique was found to be suitable for the analysis of trace components of food aromas, especially for the analysis of relatively high-boiling components.

COMPARISON OF TENAX TA AND CARBOTRAP FOR SAMPLING AND ANALYSIS

A simple method for the combined determination of adsorption and desorption capacities, as well as recovery data of selected compounds after adsorption on Tenax TA and Carbotrap, has been presented.[22]

Tenax (155 mg) was packed in a glass tube and the packed tube was cleaned under a 7 mL/min flow of helium for 110 min, followed by 10 min at 280°C. Carbotrap (170 mg) was packed in a glass tube and the packed tube was cleaned under a 7 mL/min flow of helium for 50 min, followed by 10 min at 360°C.

Two typical representatives were chosen from various chemical groups which may have been found in outdoor air, have been measured in indoor air, or were expected to be emitted from known sources. n-Decane was used as an internal standard. The test compound and the internal standard in 0.50 μL of a solvent (acetone, ethanol, hexane, isooctane, or toluene) were applied to a silanized glass wool plug preceding the adsorbent in a glass tube. A personal pump was used to draw the specimen through the sorbent with 100 mL/min of air. Total air sampling volumes of 1.0 L ± 5% and 5.0 L ± 5% were used.

The samples were thermally desorbed, separated by gas chromatography, and detected with a flame ionization detector. For each single test compound, sampling and analysis were perfomed three times for the 1-L and the 5-L volumes. Reference values were determined three times and blanks were determined two times.

The results showed that:

1. Many volatile organic compounds found in trace amounts in air could be determined quantitatively by adsorption on Tenax TA and Carbotrap, and by subsequent thermal desorption.
2. Tenax TA seemed to be inert towards the compounds tested; α-pinene and aldehydes showed some reactivity on Carbotrap.
3. Higher-boiling-point (up to 270°C) compounds could be quantitatively desorbed from Tenax, but not from Carbotrap.
4. Substantial losses, probably due to breakthrough, were observed on both adsorbents for very volatile organic compounds and various polar volatile organic compounds.

APPARENT REACTION PRODUCTS DESORBED FROM TENAX

Distributed air volume sets of tandem Tenax GC beds were collected to assess the effects of artifact reactions on data from air sampling.[23]

Tenax GC, of differing and uncontrolled histories, were solvent cleaned and packed in cartridges containing 1.2-g beds. Ambient air was sampled, and then the Tenax GC was thermally desorbed and the eluent was cryofocused on a glass capillary in a gas chromatograph-mass spectrometer. Detection limits for the substances reported in the study were between 1 and 3 ng.

Sampling of ambient air using Tenax GC was found to be practical. However, there were complications. Retention volumes sometimes differed greatly from literature values. Chemical reactions during sampling and thermal desorption were found to be common. Corrupted behavior was always displayed by benzaldehyde, 1-phenylethanone, and benzonitrile; use of Tenax GC to determine atmospheric concentrations for these compounds was considered to be unlikely to succeed. Inconsistent patterns of behavior were observed for aromatic and halogenated hydrocarbons; for these classes, therefore, an empirical evaluation of data using distributed air volume sets seemed necessary.

It was concluded that empirical evaluation of data from Tenax GC sampling of ambient air was warranted in each sampling situation.

FIELD AUDIT RESULTS WITH ORGANIC GAS STANDARDS

Two field audits of ambient sampling systems using Tenax GC were described.[24] The audits were conducted with National Bureau of Standards (NBS, now NIST) volatile organic compound (VOC) reference standards.

The audits provided an assessment of the accuracy and precision of an EPA-operated ambient air sampler that used Tenax GC to collect VOCs from air. The audits also provided information, useful for those considering the use of Tenax GC for VOCs, on the capabilities of Tenax GC under field sampling conditions.

The entire air monitoring system was audited on-site in the field. Samples were returned to the laboratory, at the completion of the audit, where they were analyzed by GC-MS-COMP. The Tenax GC cartridges were anaylzed by desorption at 270°C for 8 min into the GC-MS-COMP.

The Tenax GC was cleaned by Soxhlet extraction with methanol and pentane (8 hours for each). After removal of the solvent, the sorbent was also thermally desorbed at 270°C for 4 hours.

The results of the field audits identified some problems with Tenax GC-equipped samplers. However, the audits showed that successful field evaluations of air monitoring systems using Tenax GC could be conducted with gas cylinders containing volatile organic compounds. Traceability to a common reference can be established by the use of cylinders certified or analyzed by NIST.

SAMPLING AND THERMAL DESORPTION EFFICIENCY OF TUBE-TYPE DIFFUSIVE SAMPLERS

The development of a simple experimental procedure for the selection of proper adsorbent/analyte combinations for tube-type diffusive samplers has been presented.[25] Characteristics of experimentally determined adsorption isotherms were used to develop a straightforward procedure for finding the optimal adsorbent to be used in thermally desorbed diffusive samplers for a given chemical compound.

The adsorbents used in the study were

1. Tenax TA
2. Chromosorb 106
3. Spherocarb

A specially built dynamic air dilution system was used to generate test atmospheres in the concentration range of 1 to 1000 ppm. The adsorbates used and the adsorbent with which they were used were

1. Styrene-Tenax TA
2. Benzene-Tenax TA

3. 1,3-Butadiene-Tenax TA
4. 1,3-Butadiene-Spherocarb
5. Benzene-Chromosorb 106

The adsorbents were packed in standard adsorption tubes. The tubes packed with Tenax TA were exposed to the adsorbates at different concentrations for different lengths of time. The adsorbents were thermally desorbed by a two-stage automatic thermal desorber. The desorber was coupled to a gas chromatograph fitted with a fused-silica capillary gas chromatograph column with a flame ionization detector, for analysis. In all experiments, Tenax TA was used as the adsorbent in the secondary trap of the thermal desorber.

Adsorption isotherms were determined for the three adsorbates for Tenax TA. The adsorption isotherms were plots of mass of adsorbate adsorbed, in $\mu g/cm^3$, against concentration in ppm. Experimental uptake rates of adsorbate/adsorbent combinations that had isotherms in the indicated safe area were found to differ from ideal uptake rates by less than 10%. Combinations chosen on this basis could be applied for exposure doses up to at least 20,000 ppm/min.

DETERMINATION OF VOLATILE AND SEMIVOLATILE MUTAGENS IN AIR

Tenax TA was one of the adsorbents (the others were charcoal, Carbosieve SIII, XAD-4, and Chromosorb 102) used in a comparison of the trapping efficiency of some solid-phase adsorbents for the volatile mutagenic compounds:[26]

1. Dichloromethane
2. Ethylene dibromide
3. 4-Nitrobiphenyl
4. 2-Nitrofluorene
5. Fluoranthene

The compounds were used to optimize collection and extraction methods for air samples.

In preparation for air sampling, Tenax TA was heated to 230°C and purged with helium. Tenax TA was used to evaluate trapping of 4-nitrobiphenyl, 2-nitrofluorene, and fluoranthene. The adsorbates were recovered by solvent extraction using ethyl acetate.

Tenax TA trapped 4-nitrophenyl effectively. It trapped 2-nitrofluorene less effectively than did XAD-4. Tenax TA and Chromosorb 102 were similar in their abilities to trap 4-nitrobiphenyl, 2-nitrofluorene, and fluoranthene from an airstream.

PERFORMANCE OF A TUBE-TYPE DIFFUSIVE SAMPLER

Tenax GC was used in a study of the performance of a tube-type sampler for organic vapors in air.[27]

The sampler had a low sampling rate of about 1 ng/ppm/min. For benzene sampled in a Tenax GC tube with a 2.3-cm air gap, the uptake rate was 1.4 ± 0.06 ng/ppm/min over the range 1.3–78.6 ppm (2 hour exposure); for 50 ppm of benzene sampled on Tenax GC as above, the uptake rate was 1.4 ± 0.04 ng/ppm/min over the range 13–95% relative humidity at 20°C.

The authors observed a decrease in uptake with time of exposure, which they attributed (for Tenax GC, Porapak, and XAD-2) to a small vapor pressure gradient at the adsorbent surface which would reduce the concentration gradient of pollutant and, hence, the rate of adsorption would be reduced. They did not consider the observed change to be very serious. The moderate effect observed with benzene sampled on Tenax GC might be reduced by substituting a stronger adsorbent such as Porapak N.

PERFORMANCE OF A TENAX-GC ADSORBENT TUBE

An analysis was made of the adsorption characteristics of Tenax GC.[28]

The adsorbent tube used in the studies was constructed of stainless steel tubing (75 mm × 4.5 mm I.D.) and contained 0.13 ± 0.01 g of 40/60 mesh Tenax GC. Before use, the tubes were conditioned under nitrogen at 250°C for 16 hours. The following topics were investigated:

1. Determination of safe sampling volumes (defined as 50% of the measured retention volume)
2. Direct and indirect measurement of breakthrough volume
3. Extrapolation of retention volume data
4. Influence of flow rate on breakthrough volumes
5. The effects of temperature and humidity on breakthrough volumes
6. Influence of vapor concentration on breakthrough volumes

Temperature markedly affected breakthrough; breakthrough was virtually independent of humidity.

EVALUATION OF CHROMATOGRAPHIC SORBENTS

Tenax GC was one of the chromatographic sorbents used in air pollution studies that were evaluated in laboratory and field experiments.[29]

The laboratory and field experiments were performed to permit detection of the potential for *in situ* formation of chemical substances or their decomposition on the surface of Tenax GC, XAD-2, and carbon sorbents.

Tenax GC was purified by Soxhlet extraction with methanol and *n*-pentane, respectively, for 18 hours. After drying under a nitrogen atmosphere, the sorbent was heated to 150°C for 2 hours in a vacuum oven, sized into a 35 to 60 mesh range, and packed into glass tubes. The sample cartridges were preconditioned by heating to 275°C for 20 min under a helium purge of 20 to 30 mL/min. After cooling in precleaned culture tubes, the containers were sealed to prevent contamination of the cartridges.

Replicate samples and blanks were analyzed by gas chromatography equipped with flame ionization detection. Vapors were transferred from the cartridge sampler to the analytical system by thermal desorption. A gas-liquid chromatograph (GLC) with electron-capture detection and 0.2% Carbowax 1500 on a Carbopack C column was also used. A GLC-mass spectrometer-computer system was used for analyzing Tenax GC cartridges when structural confirmation was required.

Halogenated hydrocarbons, in trace quantities, were produced in reactions between bromine, chlorine, 2-butene, and cyclohexene, but not with ethylene or propylene. These *in situ* reactions were prevented by using a glass fiber filter impregnated with sodium thiosulfate in front of the sorbent cartridge to quench the reactive gases.

SYNTHESIS AND CHARACTERIZATION OF POROUS POLYIMIDES FOR AIR SAMPLING

Tenax GC was used as a reference sorbent in a study of the synthesis and characterization of porous polyimides for air sampling of volatile organic compounds.[30]

A range of aromatic polyimide resins were synthesized and evaluated as sorbents for vapor phase organic analysis, as an alternative to commercially available materials. A total of 55 different polyimide resins was prepared. Laboratory tests performed to evaluate the acceptability of each resin as sorbent for trace level organic analysis included: (1) thermogravimetric analysis, (2) gas chromatographic analysis of sorbent background, and (3) retention volume measurements by temperature-programmed gas chromatography.

Sorption properties of the four most promising polyimides, and Tenax GC as a reference sorbent, were further investigated by isothermal gas chromatography measurements at zero surface coverage.

Results for the selected polyimide sorbents in tests related to the trace analysis of organics in air were found to be comparable to or better than the corresponding values for Tenax GC.

COLLECTION AND ANALYSIS OF HAZARDOUS ORGANIC EMISSIONS

Tenax was used in a research program, the primary goal of which was the development of methodology for the reliable and accurate collection and analysis of hazardous vapors present simultaneously in the atmosphere, down to nanograms per cubic meter amounts.[31]

The system under development encompassed the collection and concentration of organic pollutants from ambient air using cartridges, tubes packed with polymeric beads. After sampling, the cartridges were thermally heated under a helium flow. The adsorbed compounds were desorbed, cryofocused, and subsequently introduced into a high-resolution glass or fused capillary for characterization and measurement by GC-MS-COMP techniques.

A 1.5 cm × 6.0 cm bed of Tenax GC (35/60 mesh) was used in the cartridges to concentrate organic vapors. Two or three cartridges were designated as blanks for assessing occurrence of contamination. The thermal desorption of the sampling cartridges was accomplished by passing helium through the cartridge in a chamber preheated to about 270°C, at a flow rate of about 15 mL/min. The effluent from the cartridge passed into a liquid nitrogen-cooled trap.

THERMAL DESORBABILITY OF POLYCYCLIC AROMATIC HYDROCARBONS AND PESTICIDES FROM TENAX GC

The desorbability of a set of compounds which were generally not considered to be easily thermally desorbable from Tenax GC was investigated.[32]

A main objective of the research was to investigate the extent to which increasing the carrier flow rate could be used to accomplish quantitative transfer to a fused silica capillary column at minimal desorption temperatures while maintaining high resolution in subsequent chromatographic runs.

The desorbabilities of polycyclic aromatic hydrocarbons (PAHs) were investigated as a function of helium carrier gas flow rate and desorption temperature. Other carrier gases of larger molecule size than helium were also investigated.

Cartridges were packed with 0.110 g of 35/60 mesh Tenax GC. An injection septum was mounted on the top end of the cartridge and the bottom end of the cartridge was interfaced to a capillary column. In each experiment, 4.0 μL of a standard was loaded on top of the column. The loading of the standard was followed by a series of eight 20-min desorptions and eight accompanying runs. Carbon dioxide, methane, ethane, and propane were found to give more facile desorption than helium at the same flow rate for the study compounds. The desorbed compounds were trapped on a fused silica capillary column at −30°C.

Despite the use of high flow rates for desorption carrier gas, excellent resolution and separation number performance was maintained.

CHEMICAL TRANSFORMATIONS DURING AMBIENT AIR SAMPLING FOR ORGANIC VAPORS

A sorbent collection method using Tenax GC was central to a study of potential halogenation, nitrosation, and ozonization reactions and their inhibition with sodium thiosulfate during the collection of vapor-phase organics from ambient air.[33] A thermal desorption capillary gas chromatography-mass spectrometry (GC-MS) method was used to analyze analytes collected on a Tenax GC sampling cartridge.

By sampling ambient air using Tenax GC cartridges and Tenax GC spiked with deuterated compounds, field *in situ* reaction studies were conducted. Subsequently, the cartridges were analyzed to detect and identify several deuterated oxidation and halogenated products.

Oxidation was prevented when filters employed for removing particulates were impregnated with 5 to 10 mg of sodium thiosulfate and placed in front of the sorbent cartridge; halogenation reactions were also considerably reduced.

TRACE ORGANIC VAPOR POLLUTANTS IN AMBIENT ATMOSPHERES

Tenax GC was among several sorbents evaluated as collection media for the quantitative concentration and analysis of volatile, hazardous, vapor-phase compounds from ambient atmospheres under a variety of conditions relevant to field sampling.[34]

The sorbents evaluated were

1. Graphitized carbon
2. PBL carbon
3. PCB carbon
4. SAL9190
5. M180B
6. Tenax GC (35/60 mesh)
7. Porapak Q (100/120 mesh)
8. Chromosorb 101 (60/80 mesh)
9. Chromosorb 102 (60/80 mesh)
10. Chromosorb 104E (60/80 mesh)

Tenax GC was selected as the best available sorbent for general use because of its (1) wide range of applicability, (2) thermal stability, and (3) low retentive index for water.

EVALUATION OF SAMPLING METHODS FOR GASEOUS ATMOSPHERIC SAMPLES

Tenax GC was used in a research program conducted to test and evaluate several alternatives for collecting and transferring samples to the laboratory for the analysis of a variety of toxic organic pollutants by gas chromatography.[35]

Sample storage media included:

1. Three types of polymeric bags
2. Glass bulbs

3. Electropolished canisters
4. SUMMA polished canisters
5. Tenax GC cartridges
6. Charcoal cartridges
7. Nickel cryogenic traps

A total of 27 compounds, including the following, were used to test the storage media:

1. Hydrocarbons
2. Aromatics
3. Halogenated hydrocarbons
4. Halogenated aromatics
5. Oxygen-containing compounds
6. Nitrogen-containing compounds
7. Sulfur-containing compounds

The potential effect of inorganic gases as interferences during the collection of test compounds was quantitatively studied. An automatic two-channel ambient air sampler using sorbent cartridges as the collection medium was designed and fabricated. A quality control and quality assurance program was established and maintained for all measured and analyzed data.

SAMPLING CARTRIDGE FOR COLLECTION OF ORGANIC VAPORS

A new (at that time) Tenax-filled sampling cartridge has been designed;[36] the design, manufacturing procedure, and preliminary test results were given.

The cartridge incorporated the standard glass sampling tube that sealed the sorbent tube from external contamination. The sampling cartridges were assembled in a clean laboratory area, the sealing was checked by pressurizing the cartridge with prepurified nitrogen and monitoring the pressure. The cartridge was then brought to a positive pressure of 1 atm and stored. The sampling cartridges were taken, as needed, to the field in a pressurized condition.

TOXIC AIR POLLUTANTS MEASURED FOR 355 PERSONS

Cartridges containing Tenax were used to measure toxic, carcinogenic, or mutagenic organic compounds in a second phase of E.P.A.'s Total Exposure Assessment Methodology study.[37]

Cartridges containing about 2 g of 40/60 mesh purified Tenax were used to collect personal and outdoor air samples during the normal daily activities of participants for 12-hour periods. The cartridges were subsequently analyzed by thermal desorption and cryofocusing of the organic vapors followed by capillary gas chromatography-mass spectrometry-computer analysis. About 3000 air samples were collected, of which 1000 were quality control samples.

PERSONAL EXPOSURE TO VOLATILE ORGANIC COMPOUNDS

Tenax GC was used in a pilot study to test methods of estimating personal exposures to 20 target toxic substances and corresponding body burdens.[38]

One of the objectives of the volatile organics portion of the study was to test a personal monitor employing Tenax GC adsorbent to collect selected organic compounds from breathing-zone air for 5 to 10 hours. Target compounds were selected on the basis of their extensive use and likely toxicity. All air exposures were measured using personal air quality monitors. All volatile organics except vinyl chloride were collected using glass tubes containing 1.5 g of Tenax GC.

Samples were analyzed using thermal desorption and purging by helium into a liquid-nitrogen-cooled nickel capillary cryogenic trap, followed by high-resolution glass column gas chromatography-mass spectrometry techniques.

About 230 personal air samples and 66 breath samples were analyzed for the target chemicals.

INFLUENCE OF PERSONAL ACTIVITIES ON EXPOSURE TO VOLATILE ORGANIC COMPOUNDS

In a study carried out to determine the effects of each of approximately 25 activities on personal exposure, indoor air concentrations, and exhaled breath, Tenax GC was used to collect personal, indoor, and air samples.[39]

Approximately 9 L of air was pulled through a bed of Tenax GC contained in a glass tube, to collect volatile organic compounds. Analysis of the collected compounds was by thermal desorption at 260°C with a nominal helium flow into a liquid-nitrogen-cooled nickel capillary trap. The condensed vapors from the trap were then introduced into a high-resolution fused silica capillary chromatography column by ballistic heating of the nickel trap at 250°C. Electron impact mass spectrometry was used to identify and quantitate sample constituents, by measuring the intensity of the extracted ion current profile.

The study showed that about 20 common activities can result in sharply increasing personal exposures to about 15 toxic volatile organic compounds.

CAPILLARY CHROMATOGRAPHIC ANALYSIS OF VOLATILE ORGANIC COMPOUNDS

Tenax GC has been used in applications of capillary GC columns for the analysis of volatile organic compounds (VOCs) in the indoor environment.[40]

Solid sorbent tubes were prepared. They consisted of 180 mm × 4 mm I.D. glass tubes into which 200 mg of Tenax GC (60/80 mesh) was packed. The ends of the tubes were plugged with glass wool.

Prior to being packed in the tubes, the Tenax GC was Soxhlet extracted with methanol and hexane for 24 hours each. The packed tubes were conditioned at 270°C for a minimum of 18 hours while being flushed with a stream of nitrogen with a flow rate of 20 mL/min. The tubes were then cooled to room temperature while flushing with nitrogen continued. The cooled tubes were placed in clean aluminum O-ring-sealed containers for transportation to (and from) sampling sites. The solid sorbent tubes were used for sampling of ambient VOCs in controlled environmental chambers and office buildings.

The VOCs collected in the solid sorbent tubes were thermally desorbed and analyzed by gas chromatography-mass spectrometry.

POLYCYCLIC AROMATIC HYDROCARBON AND NITROARENE CONCENTRATIONS

Tenax GC cartridges were used for sampling in the measurement of polycyclic aromatic hydrocarbons (PAHs) and nitroarenes during a wintertime, high-NO_x episode at a location in Southern California.[41]

Tenax GC cartridges were prepared consisting of 10 cm × 4 mm I.D. Pyrex tubes packed with 0.1 g of Tenax GC and doped with 610 ng of naphthalene-d_8. When sampling ambient air, the flow rate through the cartridges was 1.15 L/min, yielding about 0.8 m^3 volume sampled for each 12-hour sampling period. Downstream of the first cartridge, a second cartridge was placed to check for breakthrough from the first cartridge.

SAMPLING AND ANALYSIS OF ORGANIC COMPOUNDS OF CHEMICAL DEFENSE INTEREST IN AIR

In the development of an integrated approach to sampling and analysis, capable of tracking rapidly changing concentrations of volatile organics in air, Tenax TA was used for sampling.[42]

Integration of field sampling and laboratory analysis through a common sampling component was the principal objective for the system. The common sampling component was the sample carousel. The sample carousel was a 50-sample tube holder that was loaded into an air sampler for collection of samples. After completion of sampling, the sample carousel was brought back to the laboratory and inserted into an automated thermal desorption unit for gas chromatographic analysis.

In preparing solid adsorbent sampling tubes, borosilicate glass minitubes (38 mm × 2 mm I.D.) were packed with approximately 14 mg of Tenax TA. The adsorbent was held in place by silanized glass wool or stainless steel screens. During the packing of the minitubes, the pressure drop across the minitubes was monitored to meet the specifications of the pumping system of the air sampler. Then, 50 mL of air above a vessel suspected of containing a chemical warfare agent was drawn through a minitube packed with Tenax TA, in order to evaluate the system under field conditions. The sample was taken back to the laboratory for thermal desorption-gas chromatography analysis with flame photometric detection. The thermal desorption oven temperature was 200 or 220°C.

The system was field tested and was successfully used to collect and analyze an unknown vapor sample containing the chemical warfare agent, mustard.

EVALUATION OF TENAX GC AND XAD-2 AS POLYMER ADSORBENTS

Tenax GC and XAD-2 were evaluated as polymer adsorbents for sampling fossil fuel combustion products containing nitrogen oxides.[43]

Before use, the Tenax GC was Soxhlet extracted for 24 hours with glass-distilled *n*-pentane, to extract traces of residue, and then dried in an oven at 110°C. The adsorbents were packed in 15 cm × 2.5 cm I.D. brass tubes which were placed in series. At each end of the tube, stainless steel screens were placed to keep the adsorbent in place. Two other samplers were placed behind the first sampler to collect any organics that passed through the first adsorbent sampler.

Diluted stack gases at a flow rate of 10 mL/min were sampled for from 36 to 103 min. The concentration of NO_x in the undiluted effluent ranged from 460 to 810 ppm. A gas chromatograph-mass spectrometer was used to make qualitative identifications of compounds.

The major NO_2-decomposition products of Tenax GC were 2,6-diphenyl-*p*-quinone and diphenylquinols, which are not mutagenic and did not interfere with the gas chromatographic elution and analysis of vapor-phase organics adsorbed on the polymer. Greater than 90% of the gas chromatographable vapor-phase organics were collected on the first two stages of a three-stage adsorbent sampler filled with Tenax GC.

The results of the evaluation indicated that Tenax GC was more suitable than XAD-2 for sampling vapor-phase organics in combustion effluents containing oxides of nitrogen.

IN SITU DECOMPOSITION PRODUCT ISOLATED FROM TENAX GC

The effect on Tenax GC of the polymer degradation that leads to formation of 2,6-diphenyl-*p*-quinone (DPQ) was investigated.[1] The investigation involved comparisons of the performance of Tenax GC as a gas chromatographic column material.

The Tenax GC used was thoroughly extracted (24 hours in a Soxhlet extractor) with pentane and then methanol. The adsorbent was then dried and "activated" overnight at 200°C.

Two 10 ft × 2 mm I.D. glass GC columns were packed with Tenax GC, one with the material subjected to stack sampling conditions that form DPQ and one with the cleaned "untreated" material.

The authors concluded that the degradation of Tenax GC does not affect the efficiency or capacity of a sampling system which may utilize this adsorbent.

FORMATION OF METHYLNITRONAPHTHALENES

Tenax GC cartridges were used in an investigation of the nitro isomers formed from the gas-phase reactions of 1- and 2-methylnaphthalene with the OH radical (in the presence of NO_x) and with N_2O_5.[44]

The gas-phase reactions were carried out in a 6400-L all-Teflon environmental chamber equipped with black-light irradiation. The gas-phase concentrations of the methylnaphthalenes (volatile polycyclic hydrocarbons, PAHs) were monitored, before and after the gas-phase reactions, by gas chromatography with flame ionization detection. A 100-mL sample from the chamber was pulled through a Tenax GC cartridge. The cartridge was thermally desorbed at 250°C onto the head of a gas chromatography column. After introduction and mixing of the reactants, and after the lights were turned on at maximum intensity for 1.5 min, a 100-mL cartridge sample was used to quantify as above the remaining methylnaphthalene. The same procedure was used after N_2O_5 exposures.

EMISSIONS OF PERCHLOROETHYLENE FROM DRY-CLEANED FABRICS

Tenax sorbent was used in a study conducted to evaluate the emissions of perchloroethylene (tetrachloroethylene) from dry-cleaned fabrics.[45]

Environmental test chambers were used to study the loss of perchloroethylene from dry-cleaned fabrics. Fabrics were cut to size within 1 hour of being picked up at the dry cleaners. The cut fabric sections were hung on wire racks and placed in the test chambers.

Tandem glass cartridges filled with Tenax and Tenax/charcoal sorbents were used for collection of samples. A portion of the chamber air stream was pulled through the cartridges at a flow rate of 0.1 L/min. Sample volumes of 0.5 to 10 L were collected during a sampling time of from 5 to 100 min.

The cartridges were thermally desorbed at 220°C. The desorbed compounds from the cartridges were delivered to the Tenax/charcoal concentrator column of a purge-and-trap unit. The collected compounds on the concentrator column were desorbed to the analytical column of a gas chromatograph by rapidly heating the concentrator column. Perchloroethylene was identified by gas chromatograph retention time and quantified by the response of a flame ionization detector.

DETERMINATION OF NICOTINE IN INDOOR ENVIRONMENTS USING A THERMAL DESORPTION METHOD

Tenax GC was used as the adsorption material in the development, evaluation in controlled environmental tobacco smoke atmospheres in chambers and offices, and field validation of a thermal-desorption-based personal monitoring system for nicotine.[46]

Air-sampling cartridges consisted of 16 cm × 1/4 in. O. D. treated borosilicate glass tubing packed with 200 mg of 35/60 mesh Tenax GC. The packed cartridges were conditioned at 250°C before use. After sampling by pumping air at a constant rate through the cartridges, the cartridges were analyzed by triethylamine-assisted thermal desorption gas chromatography with nitrogen-selective detection. Collection efficiencies and desorption efficiencies for the cartridges were determined.

The personal monitoring system was evaluated in controlled-atmosphere chambers, at a variety of work sites, and in 36 restaurants. Measured concentrations of nicotine ranged from 0.5 to 37.2 μg/m³.

Results obtained with the Tenax GC method for nicotine concentrations in restaurants were comparable to those obtained by researchers using different methods for sampling in restaurants and other public places.

TENAX TRAPPING OF GAS PHASE ORGANIC COMPOUNDS IN ULTRA-LOW-TAR CIGARETTE SMOKE

Tenax GC was used in a method for the analysis of gas phase organic compounds in ultra-low-tar delivery cigarette smoke.[47]

Cigarettes were smoked directly through a filter and a Tenax GC trap arranged in series. The components of the smoke collected on the Tenax GC were analyzed by thermal-desorption programmed-temperature gas chromatography. External standards were used to quantitate the components.

The 35/60 mesh Tenax GC was treated initially with successive batch extractions (two each) with equal or greater volumes of water, 50% methanol, methanol, ether, and pentane. After the last pentane wash, the Tenax GC was transferred to a large Soxhlet extractor for continuous extraction with n-pentane for 48 hours. The residual pentane was blown off with a nitrogen sweep and then the Tenax GC was conditioned in a 1-L stainless steel bomb under helium flowing at 300 mL/min for 1 hour at room temperature, followed by temperature programming at 2°C/min to 250°C, where it was held for 24 hours. The Tenax GC was cooled to room temperature under flowing helium and was then transferred in a clean air box to a brown bottle having an aluminum foil-lined cap.

The Tenax GC traps consisted of 11 cm × 9 mm O.D. × 5 mm I.D. heavy-wall Pyrex desorption tubes, tapered at one end to fit the end of a gas chromatograph injector port, and filled with 220 ± 15 mg of Tenax GC held in place by solvent-washed and thermally conditioned glass wool.

REFERENCES

1. Neher, M. B., and P. W. Jones. *"In Situ* Decomposition Product Isolated from Tenax GC While Sampling Stack Gases," *Anal. Chem.* 49:512–513 (1977).
2. van Wijk, R. *J. Chromatogr. Sci.* 8:418–419 (1970).
3. Janak, J., J. Ruzickova, and J. Novak. *J. Chromatogr.* 99:689–696 (1974).
4. Bertsch, W., R. C. Chang, and A. Zlatkis. *J. Chromatogr. Sci.* 12:175 (1974).
5. Boyko, A. L., M. E. Morgan, and L. M. Libbey, in *Analysis of Foods and Beverages: Headspace Techniques,* G. Charalambous, Ed. (London: Academic Press, 1978) p. 57.
6. Cole, R. A. *J. Sci. Food Agric.* 31:1242 (1980).
7. Sakodynskii, K., L. Panina, and N. Klinskaya. *Chromatographia* 7:339 (1974).
8. Bertsch, W., E. Anderson, and G. Holzer. *J. Chromatogr.* 112:701 (1975).
9. Barnes, R. D., L. M. Law, and A. J. MacLeod. *Analyst (London)* 106:412 (1981).
10. MacLeod, G., and J. M. Ames. "Comparative Assessment of the Artefact Background on Thermal Desorption of Tenax GC and Tenax TA," *J. Chromatogr.* 355:393–398 (1986).
11. Holzer, G., H. Shanfield, A. Zlatkis, W. Bertsch, P. Juarez, H. Mayfield, and H. M. Liebich. "Collection and Analysis of Trace Organic Emissions From Natural Sources," *J. Chromatogr.* 142:755–764 (1977).
12. Venema, A., N. Kampstra, and J. T. Sukkel. "Investigation of the Possible Oxidation by Ozone of Organic Substrates Adsorbed on Tenax," *J. Chromatogr.* 269:179–182 (1983).
13. *Chrompack News Special,* Tenax. Middleburg, 1982, p. 1.
14. Rothweiler, H., P. A. Wager, and C. Schlatter. "Comparison of Tenax TA and Carbotrap for Sampling and Analysis of Volatile Compounds in Air," *Atmos. Environ.* 25B:231–235 (1991).
15. Figge, K., A. M. Dommrose, W. Rabel, and W. Zerhau. "Sammelund Analysensystem zur Bestimmung organisher Spurenstoffe in der Atmosphare," *Fres. A. Anal. Chem.* 327:279–292 (1987).
16. De Bortoli, M., H. Knoppel, E. Pecchio, and H. Vissers. "Performance of a Thermally Desorbable Diffusion Sampler for Personal and Indoor Air Monitoring," in *Indoor Air '87,* Eds. B. Seifert, H. Esdorn, M. Fischer, Ruden, and J. Wegner, (Berlin:Inst. Water, Soil Air Hyg., 1987) pp. 139–143.
17. Class, T., and K. Ballschmitter. "Chemistry of Organic Traces in Air. VI. Distribution of Chlorinated C_1–C_4Hydrocarbons in Air Over the Northern and Southern Atlantic Ocean," *Chemosphere* 15:413–427 (1986).
18. Pellizzari, E., B. Demain, and K. Krost. "Sampling of Organic Compounds in the Presence of Reactive Inorganic Gases with Tenax GC," *Anal. Chem.* 56:793–798 (1984).
19. Clark, A. I., A. E. McIntyre, N. J. Lester, and R. J. Perry. *J. Chromatogr.* 252:147–157 (1982).

20.   Pellizzari, E. D. *Environ. Sci. Technol.* 16:781–785 (1982).

21.   MacLeod, G., and J. M. Ames. "Comparative Assessment of the Artefact Background on Thermal Desorption of Tenax GC and Tenax TA," *J. Chromatogr.* 355:393–398 (1986).

22.   Rothweiler, H., P. A. Wager, and C. Schlatter. "Comparison of Tenax TA and Carbotrap for Sampling and Analysis of Volatile Organic Compounds in Air," *Atmos. Environ.* 25B:231–235 (1991).

23.   Walling, J.F., J. E. Bumgarner, D. J. Driscoll, C. M. Morris, A. E. Riley, and L. H. Wright. "Apparent Reaction Products Desorbed from Tenax used to Sample Ambient Air," *Atmos. Environ.* 20:51–57 (1986).

24.   Crist, H. L., and W. J. Mitchell. "Field Audit Results with Organic Gas Standards on Volatile Organic Ambient Air Samplers Equipped with Tenax GC," *Environ. Sci. Technol.* 20:1260–1262 (1986).

25.   Van den Hoed, N., and M. T. H. Halmans. "Sampling and Thermal Desorption Efficiency of Tube-Type Diffusive Samplers: Selection and Performance of Adsorbents," *Am. Ind. Hyg. Assoc. J.* 48:364–373 (1987).

26.   Wong, J. M., N. Y. Kado, P. A. Kuzmicky, H.-S. Ning, J. E. Woodrow, D. P. H. Hsieh, and J. N. Seiber. "Determination of Volatile and Semivolatile Mutagens in Air Using Solid Adsorbents and Supercritical Fluid Extraction," *Anal. Chem.* 63:1644–1650 (1991).

27.   Brown, R. H., and K. T. Walkin. "Performance of a Tube-Type Diffusive Sampler for Organic Vapours in Air," Anal. Proc. Fifth Int. SAC Conf., May, 1981, pp. 205–208.

28.   Brown, R. H., and Purnell, C. J. "Collection of Trace Organic Vapour Pollutants in Ambient Atmospheres; The Performance of a Tenax-GC Adsorbent Tube," *J. Chromatogr.* 178:79–90 (1979).

29.   Bunch, J. E., and E. D. Pellizzari. "Evaluation of Chromatographic Sorbents Used in Air Pollution Studies," *J. Chromatogr.* 186:811–829 (1979).

30.   Demain, B., K. Lam, A. Schindler, and E. D. Pellizzari. "Synthesis and Characterization of Porous Polyimides for Air Sampling of Volatile Organic Compounds," in *Identification and Analysis of Organic Pollutants in Air,* L. H. Keith, Ed. (Boston: Butterworth Publishers, 1984), pp. 95–124.

31.   Krost, K. J., E. D. Pellizzari, S. G. Walburn, and S. A. Hubbard. "Collection and Analysis of Hazardous Organic Emissions," *Anal. Chem.* 54:810–817 (1982).

32.   Pankow, J. F., and T. J. Kristensen. "Effects of Flow-Rate and Temperature on Thermal Desorbability of Polycyclic Aromatic Hydrocarbons and Pesticides from Tenax-GC," *Anal. Chem.* 55:2187–2192 (1983).

33.   Pellizzari, E. D., and K. J. Krost. "Chemical Transformations during Ambient Air Sampling for Organic Vapors," *Anal. Chem.* 56:1813–1819 (1984).

34.   Pellizzari, E. D., J. E. Bunch, R. E. Berkley, and J. McRae. "Collection and Analysis of Trace Organic Vapor Pollutants in Ambient Atmospheres. The Performance of a Tenax GC Cartridge Sampler for Hazardous Vapors," *Anal. Lett.*9:45–63 (1976).

35.   Pellizzari, E. D., W. F. Gutknecht, S. Cooper, and D. Hardison. "Evaluation of Sampling Methods for Gaseous Atmospheric Samples," U. S. Environmental Protection Agency, Final Report on Contract No. 68-02-2991.

36.   Russwurm, G. M., J. A. Stikeleather, P. M. Killough, and J. G. Windsor, Jr. "Design of a Sampling Cartridge for the Collection of Organic Vapors," *Atmos. Environ.* 15:929–931 (1981).

37.   Wallace, L. A., E. D. Pellizzari, T. D. Hartwell, C. M. Sparacino, L. S. Sheldon, and H. Zelon. "Personal Exposures, Indoor-Outdoor Relationships, and Breath Levels of Toxic Air Pollutants Measured for 355 Persons in New Jersey," *Atmos. Environ.* 19:1651–1661 (1985).

38.   Wallace, L. A., E. D. Pellizzari, T. Hartwell, M. Rosenzweig, M. D. Erickson, C. M. Sparacino, and H. Zelon. "Personal Exposure to Volatile Organic Compounds; I. Direct Measurements in Breathing Zone Air, Drinking Water, Food, and Exhaled Breath," *Environ. Res.* 35:293–319 (1984).

39.   Wallace, L. A., E. D. Pellizzari, T. D. Hartwell, V. Davis, L. C. Michael, and R. W. Whitmore. "The Influence of Personal Activities on Exposure to Volatile Organic Compounds," *Environ. Res.* 50:37–55 (1989).

40.   Bayer, C. W., and M. S. Black. "Capillary Chromatographic Analysis of Volatile Organic Compounds in the Indoor Environment," *J. Chromatogr. Sci.* 25:60–64 (1987).

41.   Arey, J., B. Zielinska, R. Atkinson, and A. M. Winer. "Polycyclic Aromatic Hydrocarbon and Nitroarene Concentrations in Ambient Air During a Wintertime High-NO Episode in the Los Angeles Basin," *Atmos. Environ.* 21:1437–1444 (1987).

42.   Hancock, J. R., J. M. McAndless, and R. P. Hicken. "A Solid Adsorbent Based System for the Sampling and Analysis of Organic Compounds in Air: An Application to Compounds of Chemical Defence Interest," *J. Chromatogr. Sci.* 29:40–45 (1991).

43.   Hanson, R. L., C. R. Clark, R. L. Carpenter, and C. H. Hobbs. "Evaluation of Tenax-GC and XAD-2 as Polymer Adsorbents for Sampling Fossil Fuel Combustion Products Containing Nitrogen Oxides," *Environ. Sci. Technol.* 15:701–705 (1981).

44.   Zielinska, B., J. Arey, R. Atkinson, and P. A. McElroy. "Formation of Methylnitronaphthalenes from the Gas-Phase Reactions of 1- and 2-Methylnaphthalene with OH Radicals and N_2O_5 and Their Occurrence in Ambient Air," *Environ. Sci. Technol.* 23:723–729 (1989).

45.   Tichenor, B. A., L. E. Sparks, M. D. Jackson, Z. Guo, M. A. Mason, C. M. Plunket, and S. A. Rasor. "Emissions of Perchloroethylene From Dry Cleaned Fabrics," *Atmos. Environ.* 24A:1219–1229 (1990).

46.   Thompson, C. V., R. A. Jenkins, and C. E. Higgins. "A Thermal Desorption Method for Determination of Nicotine in Indoor Environments," *Environ. Sci. Technol.* 23:429–435 (1989).

47.   Higgins, C. E., W. H. Griest, and G. Olerich. "Application of Tenax Trapping to Analysis of Gas Phase Organic Compounds in Ultra-Low Tar Cigarette Smoke," *J. Assoc. Off. Anal. Chem.* 66:1074–1083 (1983).

CHAPTER 3

Sorbent Tubes

INTRODUCTION

Glass tubes containing different sorbents are used extensively for collecting trace organics in the work environment. In this chapter, we review applications of sorbent tubes.

SAMPLING OF LIGHT HYDROCARBONS

Observing that the issue of sampling of light hydrocarbons had never been properly addressed in industrial hygiene, and noting that 33 hydrocarbons have 4 carbons or less, Des Tombe et al.[1] critically reviewed the literature with respect to collection and analyses of some light hydrocarbons.

The study, although general, was particularly related to the possible effects of exposure to gasoline vapor.[2-5] Among the references is a critical review of light hydrocarbon sampling methodology published by the Petroleum Association for Conservation of the Canadian Environment (PACE).

Light hydrocarbon sampling is divided into four categories by Des Tombe et al.:[1] (1) Colorimetric Detector Tubes; (2) Direct-Reading Instruments, including photoionization, catalytic combustion, infrared, and flame ionization; (3) Sampling at Ambient Temperature, using an active, passive, or grab device; and (4) Cryogenic Methods, using cold trapping in combination with gas chromatography.

Colorimetric Detector Tubes

Colorimetric tubes are nonspecific and have different sensitivities for the different hydrocarbons. Colorimetric tubes have numerous disadvantages and consequently are not generally used for accurate and precise sampling of light hydrocarbons.

Direct-Reading Instruments

Photoionization Instruments

Photoionization instruments do not measure methane, ethane, or propane and are thus not appropriate for use in monitoring light hydrocarbons.

Catalytic Combustion Instruments

Catalytic combustion instruments are not specific to any one hydrocarbon. After calibration with an appropriate hydrocarbon, total hydrocarbon with respect to the specific hydrocarbon gas will be measured.

Infrared Detection Instruments

Infrared detection instruments could give a total hydrocarbon concentration compared to a calibrating gas.

Flame Ionization Instruments

Flame ionization instruments are nonspecific but will measure total hydrocarbon after calibration with an appropriate hydrocarbon.

Ambient Temperature Devices

Sorbent Sampling

In the selection of a solid sorbent, the following should be considered:

"1. The volume of air sampled that can be passed through the sorbent without breakthrough of the compounds present
2. The degree of decomposition of sample components during preconcentration and liberation
3. Any background signal caused by the sorbent
4. The affinity of the sorbent for water
5. The simplicity, speed, and completeness of concentrated compounds" [1]

Using solid sorbents for collection of substances in air, the two basic techniques are (1) active and (2) passive.

In the active technique, air is drawn through the sorbent bed by a pump. In the passive technique, compounds diffuse into a chamber containing a solid sorbent. In both techniques, the compounds are desorbed from the sorbent by a suitable solvent or by thermal desorption.

Active Technique

The most widely used sorbent in the industrial hygiene field is activated coconut charcoal. It is recommended by the National Institute for Occupational Safety and Health (NIOSH). Activated charcoal has been used to monitor the whole range of hydrocarbons found in gasoline vapor; for the collection and analysis of the light components (C_1 to C_4) it is not effective. The subscript number refers to the carbon number, the number of carbon atoms in a compound. Graphitized carbon black and carbon molecular sieves (CMS) have been used to sample airborne contaminants. Graphitized carbon black has been said to be effective for trapping from C_4 up to C_8 or heavier airborne organic compounds. Carbon molecular sieves, Carbosieves, have been recommended to collect C_1 to C_4 compounds and have been effective in trapping C_2 to C_4 compounds. An adsorbent tube, properly designed, can effectively collect C_2 compounds up to about C_{10} to C_{12}.

Passive Technique

In the passive technique, vapor is transported to the adsorbent by diffusion. As sampling proceeds, the loss of light hydrocarbons may counteract the sampling process. Deviations from Fick's first law of diffusion have been considered to be more severe with the light hydrocarbons. For these and other reasons, it has been concluded that passive sampling is not well suited for volatile light hydrocarbons.

Grab Sampling Techniques

Grab sampling devices include:

1. Gas bags
2. Glass vessels
3. Syringes
4. Canisters

The grab sample is simple and convenient, although for use for light volatile components it requires a preconcentration step.

Gas bags: a variety of plastic bags has been used, including

1. Aluminized
2. Halar
3. Mylar
4. Saran

5. Tedlar
6. Teflon

From such bags there are several sources of losses:

1. Chemical reaction of the collected species with other collected species in the gas phase and with the walls of the bag
2. Imperfections in the bag
3. Permeation through the walls of the bag
4. Sorption onto the walls of the bag

Other disadvantages of gas bags include outgassing of the bag and other contamination by the bags.

SUMMA Canisters: SUMMA canisters are passivated stainless steel containers which are clean and inert. SUMMA canisters have been considered to provide very satisfactory performance for C_1 to C_{12} hydrocarbons. The canister is simple, reliable, and flexible in use.

Cryogenic Methods

Cryogenic methods are used for preconcentration of very volatile components and for solute band concentration.

Active Sampling Using a Cold Trap

In most cases, the sample is passed through a tube containing a cooled adsorbent. Although cold trap devices may be effective in retaining volatile compounds, the method is not practical for the industrial hygienist because of several associated problems.

Conclusions

"An active ambient temperature sampling device with effective adsorbents seems to be the best choice"[1] as a simple, convenient sampling method. The deficiencies of an adsorbent tube may be corrected if its effectiveness is known by comparison against the SUMMA canister (the "gold standard"). Area sampling using the volatile organic compound canister sampler (VOCCS) method may represent the best sampling technology available if further study shows that there is no adsorbent tube capable of effectively collecting light hydrocarbons.

SOLID SORBENT DETERMINATION OF HYDRAZOIC ACID IN AIR

Puskar et al.[6] developed a sampling and analytical method for the measurement of hydrazoic acid at the short-term ceiling limit (0.1 ppm) and validated the method under both laboratory and field conditions.

Hydrogen azide is released from aqueous solutions of sodium azide (sodium azoimide) and is volatile, explosive, toxic, readily escapes from the aqueous solutions, and is believed to be the "ultimate toxic agent in humans exposed to"[6] sodium azoimide.

Samples were collected on pretreated Orbo 52 activated silica gel adsorbent tubes, at a flow rate of 1.0 L/min for 15 min. The samples were stored at room temperature prior to analysis. Sample stability studies indicated that the samples would remain stable in storage for at least 4 weeks.

In preparation for sample analysis, the two sections of the adsorbent tubes were transferred to separate 2-mL vials, 1 mL of distilled water was added to each vial, each vial was capped and allowed to stand for 30 min at room temperature, and then a portion of each sample was transferred to a 300-μL microvial.

A high pressure liquid chromatograph (HPLC) with an anion-exclusion column consisting of a high-capacity ion-exchange polymer in the hydrogen ionic form was used for separation. Detection was made at 210 nm on the highly UV-absorbing azide ion.

By dissolving known amounts of sodium azide into distilled water, calibration standards were prepared.

Laboratory evaluations were performed, including:

1. Desorption efficiency
2. High-humidity sample collection
3. Refrigeration sample storage vs. room temperature sample storage
4. Sample stability over time
5. Test atmosphere sample recovery

Field evaluation in an actual sampling environment followed laboratory evaluations.

The desorption efficiency of hydrazoic acid was determined at three concentrations: 0.025, 0.1, and 1.0 ppm. These three concentrations corresponded to, respectively, 0.25×, 1×, and 10× the Occupational Safety and Health Administration (OSHA) ceiling (0.1 ppm) for a 15-min sample. The pooled desorption for the three concentrations was found to be 96%.

To determine whether refrigerated storage of spiked sample tubes was necessary, a preliminary spiking study was made. Based on the results of the study, refrigeration studies of samples was abandoned and storage by refrigeration was not recommended.

The field validation study was conducted in a facility typical of the authors' azide solution production facilities. Two types of samples were collected from six unique field atmospheres using pretreated Orbo 52 tubes, over the concentration range of less than 0.01 to 0.37 ppm. There was good agreement between field samples and spikes. At humidities as high as 82% there was no significant effect on sample recovery.

DETERMINATION OF 2,3-DIBROMOPROPANOL IN AIR

An air sampling (using sorbent sampling tubes) and gas-liquid chromato-graphic (GLC) analysis method was developed by Choudhary.[7] The sensitivity, simplicity, and rapidity of GLC recommended it for this application.

A simple glass U-tube was connected to a solid sorbent sampling tube of two sections. The sorbent was Carbotrap, 100 mg in the front section of the sampling tube and 50 mg in the backup section. Known amounts of 2,3-dibromopropanol in carbon disulfide were injected into the U-tube before air for sampling was drawn through it at 0.2 L/min for 2 to 4 hours.

Liquid spikes of the 2,3-dibromopropanol-carbon disulfide standards were injected directly on the front section of the adsorbent tubes. The spiked adsorbent tubes were either stored at room temperature or in a refrigerator for the specified period before analysis for collection efficiency or recovery studies.

The analysis utilized a gas chromatograph with the following parameters:

1. Column — 1.8 m × 2 mm I.D. ×6 mm O.D. glass column
2. Column packing — 15 SP1000 on 20/40 Carbopak C
3. Carrier gas — helium at a flow rate of 30 mL/min
4. Detector — flame ionization detector
5. Oven temperature — 180°C isothermal
6. Injector temperature — 210°C
7. Detector temperature — 250°C
8. Injection volume — 1 µL, either manually or by an autosampler

One mL of carbon disulfide was used to desorb the samples from the adsorbent tubes, which were allowed to stand for 20 to 30 min with occasional shaking before an aliquot of the solution was injected into the gas chromatograph for analysis. Since no breakthrough was found over the concentration range of the investigation, only the front sections of the air-spiked and liquid-spiked adsorbent tubes were desorbed.

The method, which used a personal pump to sample airborne 2,3-dibromopropanol onto the adsorbent and GLC-FID detection, was found to have a detection limit of 5 ng for a 1-µL injection. Recoveries by liquid spiking or vapor spiking, in the range 100 to 150 µg per sample, ranged between 90 and 100%, with an average of 96%. The residual standard deviation was 5%. There was minimal loss of samples from spiked samples stored for 3 or 4 weeks at room temperature or in the refrigerator.

The method was considered to be suitable for personal monitoring for 2,3-dibromopropanol.

USE OF ION-EXCLUSION CHROMATOGRAPHY

Hekmat and Smith[8] discussed a proposed method for the determination of low molecular weight organic acids in air which were collected on silica gel tubes, desorbed in hot (70°C) deionized distilled water, and separated by ion-exclusion chromatography using tridecafluoroheptanoic acid solution as the diluent. The proposed method was used to separate formic acid, acetic acid, propionic acid, and butyric acid.

The samples of the organic acids were drawn in air through silica gel tubes, using a controlled test atmosphere generation system, at a sampling rate of 200 mL/min. The air was drawn through the sampling media using a vacuum pump. The front section of the tubes contained approximately 570 mg of silica gel, the back section contained approximately 280 mg of silica gel. The two sections were separated by a 2-mm section of urethane foam.

The ion-exclusion chromatographic analyses were made on an ion chromatograph equipped with a separator column; 50 μL injections were made by an autosampler in the eluent stream which was being pumped at 1 mL/min. Measurements were made on a conductivity detector and recorded on an integrator. The instrument was standardized using peak height and linear mode. The mixed standards used for calibration were of three different levels.

The overall recoveries from passing room air with 50 to 55% relative humidity through spiked silica gel tubes for the four organic acids were higher than 95% for samples spiked at one-half the threshold limit value (TLV), at the TLV, and at two times the TLV, after passing laboratory air through the tubes at 200 mL/min for 2 hours. For water-saturated air, the recoveries were all above 80 %. No breakthrough was observed for acetic acid, propionic acid, and butyric acid at two times the TLV; less than 4% breakthrough was observed for formic acid. There was no change in recovery observed in samples stored for up to 7 days; desorbed samples were stable for up to 20 days.

The method was a sensitive, simple, fast, and less costly means for analysis of C_1 to C_4 fatty acids in air samples.

ADSORPTION CAPACITY MODEL FOR CHARCOAL BEDS: RELATIVE HUMIDITY EFFECTS

A simple model for humidity effects on adsorption on charcoal beds was developed and tested against published and unpublished data.[9] The assumptions made for the model are

"1. Only adsorbed (or condensed) water molecules, not water vapor molecules in the air (W), affect the capacity for adsorption of adsorbate molecules in air (A).

2. Only the capacity term ... is affected by this adsorbed water.

3. Equilibria between gas, solid, and liquid phases can be assumed for both water and adsorbate at breakthrough.

4. There exists a fixed concentration $[P]_0$ of homogeneous condensation micropores, which can contain either n molecules of water (W_nP) or m molecules of adsorbate (A_mP)."[9]

An equation was developed relating the inverse of the bed capacity to a power of the water vapor concentration, and an equation was developed relating the inverse of the ratio of a capacity or breakthrough time relative to that at a reference water vapor concentration.

The simple equilibrium model quantitatively explained observed humidity effects and it allowed extrapolation of data to untested conditions.

ADSORPTION CAPACITY MODEL FOR CHARCOAL BEDS: CHALLENGE CONCENTRATION EFFECTS

The model in the previous section was reformulated.[10] The model provided good correlations, for both dry charcoal bed data and humidity-preconditioned bed data, of effects of relative humidity and concentration of challenge vapor on the adsorption capacities of charcoal beds. The correlations could be used to interpolate and extrapolate laboratory test data to field conditions. The parameters of the model had physical significance related to the adsorbent and to the properties of the adsorbate.

THEORETICAL STUDY OF CONTAMINANT BREAKTHROUGH FOR CHARCOAL SAMPLING TUBES

Yoon and Nelson[11] applied a previously developed theoretical model[12,15] to investigate contaminant breakthrough on charcoal sampling tubes.

The theoretical model was extended to investigation of industrial hygiene sampling devices which use activated coconut-based charcoal as the adsorbent.

At a single concentration in the approximate concentration range of 200 to 500 ppm, breakthrough phenomena associated with the sampling of dichloromethane, ethyl acetate, isobutyl acetate, and perchloroethylene were studied, as were breakthrough phenomena associated with the sampling of n-heptane at several concentrations in the approximate concentration range of 100 to 1000 ppm. Calculated theoretical breakthrough curves were found to be in agreement with experimental data. It was concluded that the theory might be used to accurately predict the performance of sampling devices with coconut charcoal as the collection medium, and that the theory might be used to calculate the time of contaminant breakthrough for specific sampling problems associated with the compounds studied.

PERFORMANCE OF ACTIVATED CARBON AT HIGH RELATIVE HUMIDITIES

A simple mathematical model (extending the Dubinin-Radushkevich potential theory) was developed by Underhill[16] for the effect of relative humidity on the adsorption of water-immiscible compounds, and the predictions and accuracy of this theoretical model were examined when applied to previously published data.

An equation was developed that permitted a rapid calculation of the effect of water vapor on the adsorption of water-immiscible organic compounds. The equation used the following assumptions:

1. The adsorption potential in the presence of completely saturated air is some multiple of the adsorption potential for dry air.
2. The adsorption potential decreases linearly with a decrease in the free energy of the water vapor.
3. If the absolute value of $k_2RT/V_h[\ln(RH/100)]$ is greater than $k_1RT/V[\ln(C_s/C)]$, the organic adsorbate will displace all the water from the micropores of the adsorbent.

The factor k_2 gives the change in adsorption potential of a water-immiscible compound as a function of relative humidity (RH), R is the universal gas constant, T is the absolute temperature, V_h is the molar volume of water, ln is the natural logarithm, k_1 is a factor describing the change in the adsorption potential of a water-immiscible compound brought about by replacement of air with water in the micropores, V is the molar volume of liquified adsorbate, C_s is the concentration of the vapor of the pure adsorbate at the temperature of adsorption, and C is the ambient concentration of adsorbate. After the adsorption potential for the organic solvent has been determined, the uptake of organic adsorbate is calculated.

The theory was applied to the previously published data of Werner[17] and there was good correlation between theory and experiment. The developed equation was found to be consistent with the experimental observations that: (1) "below a certain value, the relative humidity has little effect on the uptake of the adsorbate", and (2) "the effect of relative humidity, if observed, is more severe for lower than for higher concentrations of contaminant."[16]

NEW DESORPTION TECHNIQUE FOR AIR SAMPLING SORBENT TUBES

Coyne et al.[18] evaluated a new desorption technique with several chemicals including ethylene oxide, methylene chloride, and styrene. Recoveries, optimum solvent volumes, and extraction times were evaluated. Whole tube desorption, the new desorption technique, involves adding the glass containing the sorbent to the vial containing the desorption solvents.

The following chemicals were evaluated with the technique:

1. 2,4-Dichlorophenoxy acetic acid
2. 2-Ethyl-2-oxazoline

3. Ethylene oxide
4. Methylene chloride
5. Styrene
6. 1,1,1-Trichloroethane

The recoveries for five of the six chemicals were evaluated at two known concentrations; two to three recovery determinations were made for each experiment. No distinction was made between the front and back sections of the sorbent tube. The sorbent tubes were 1-g Pittsburgh Coconut Base (PCB) Charcoal tubes, 1-g XAD-2 resin tubes, and 900-mg silica gel tubes. The desorption solvent used was carbon disulfide for all of the chemicals except for 2,4-dichlorophenoxy acetic acid, for which 50/50 acetonitrile/0.025 M phosphoric acid was used. Desorption solvent volume was either 10 or 20 mL. The extracts were shaken on a mechanical flatbed shaker for 0.5 or 2 hours. Gas chromatographic or liquid chromatographic conditions in a validated method was used for the analysis of all samples.

For several chemicals, including ethylene oxide, methylene chloride, and styrene, recovery efficiencies were comparable to those obtained using traditional solvent desorption procedures.

Advantages of the whole tube desorption method are

1. The time savings in sample preparation of sorbent tubes prior to analysis; to place a tube into a vial containing chilled solvent requires only a few seconds.
2. Sorbent losses are minimal since there is no physical transfer of sorbent.
3. In the extract there are less charcoal fines, a common cause of the clogging of chromatographic syringes.
4. Robotics can be easily implemented in the preparation of sorbent tubes for analysis.

The inability to distinguish between the front and the back sections of the conventional sorbent tube is a disadvantage of the technique. A simple modification of the design of commercially available sorbent tubes would physically separate the front and back sections.

SUPERCRITICAL FLUID EXTRACTION OF ORGANICS FROM ADSORBENTS

Dichloromethane, ethylene dibromide, 4-nitrobiphenyl, 2-nitrofluorene, and fluoranthene have been used to optimize collection and extraction methods for air samples.[19] Charcoal, Carbosieve SIII, XAD-4, Tenax TA, and Chromosorb 102 were the adsorbents tested. Supercritical fluid extraction (SFE) with carbon dioxide as the solvent was investigated for analyte recovery.

SFE can provide a more rapid and efficient extraction (when compared with conventional extraction methods), increased sensitivity, and potential sample fractionation. Supercritical fluids, at temperatures and pressures above their critical points, have densities similar to those of liquids, but with solute diffusivities and viscosities closer to those of gases. These properties result in a rapid and efficient extraction (mass transfer) of solutes. Chemical changes are minimized by the use of fluids with low critical temperatures. Carbon dioxide, the fluid of choice, has a critical temperature of 304.2 K (31.0°C), a critical pressure of 72 atm (7295 kPa), and a critical density of 0.468 g/mL.

The adsorbents used in the study were packed into 10 cm × 5 mm I.D. disposable pipets and were held in place by glass wool plugs. The plugs were precleaned by solvent extraction with ethyl acetate. Depending on the density and cost of the adsorbent, the amount of adsorbent in the pipets varied from 200 to 900 mg.

Two different systems were used to deliver model compounds to a test airstream: (1) the volatile compounds (dichloromethane and ethyl dibromide) were metered by diffusion from a reservoir through capillary tubes into a prefiltered airstream, and (2) approximately 50 µg of the semivolatile compounds (aromatics) were volatilized from the inner surface of a glass tube, heated to 60°C, upstream of the sampling tubes.

For supercritical fluid extraction, a syringe pump was used to pump supercritical carbon dioxide through an extraction cell containing adsorbent. The liquid carbon dioxide was under a helium head space pressure of 1500 psi for all extractions.

A gas chromatograph with the following parameters was used for the analysis for both dichloromethane and ethyl dibromide:

1. Column — either a 30 m × 0.53 mm I.D. J & W DB-5 megabore column with 1.5-µm film thickness, or a 30 m × 0.53 mm I.D. J & W DB-17 megabore column with 1.5-µm film thickness
2. Column flow rate — 6.8 mL/min 10% methane/argon
3. Makeup gas flow rate — 40.2 mL/min 10% methane/argon
4. Column temperature — 65°C
5. Detector — electron capture detector (ECD)
6. Detector temperature — 300°C
7. Injector temperature — 250°C

A gas chromatograph with the following parameters was used for the analysis of 4-nitrophenyl, 2-nitrofluorene, and fluoranthene:

1. Column — 30 m × 0.53 mm I.D. J & W DB-5 megabore column with 1.5-µm film thickness
2. Column flow rate — 15 mL/min helium
3. Makeup gas flow rate — 60 mL/min nitrogen
4. Column temperature — 220°C
5. Detector — electron capture detector (ECD)
6. Detector temperature — 300°C
7. Injector temperature — 250°C

Charcoal and Carbosieve SIII were the most efficient adsorbents for retaining dichloromethane. XAD-4 was the most efficient adsorbent for ethylene dibromide and the aromatic compounds. Using supercritical carbon dioxide for extraction of direct spikes of compounds, there was greater than 90% recovery of ethylene dibromide and 60 to 92% recovery of the aromatics. With ethyl dibromide and 4-nitrobiphenyl recoveries from air, integration of trapping and desorption methods with *Salmonella* microsuspension bioassay was demonstrated. Bioassay and chemical analysis gave comparable results, within 10%.

DETERMINATION OF TRACE ORGANICS IN GAS SAMPLES COLLECTED BY CANISTER

An analytical method has been developed[20] for the routine analysis of air samples collected in canisters for the target volatile organic compounds, using a gas chromatograph-mass spectrometer (GC-MS). SUMMA passivated canisters were used to collect and store whole air samples.

In the SUMMA passivated canisters, a pure chrome-nickel oxide layer is coated on the inner metal surface. This passivation layer increases the stability and the storage interval of many organic compounds and, therefore, representative air samples are yielded. Among the advantages of canister sampling are

1. Breakthrough does not occur because the actual air sample is collected.
2. Degradation products of the trapping materials do not occur.
3. Desorption efficiency of target compounds is not a problem.
4. Analysis of the canister sample can be repeated by using the remainder of the sample in the canister.
5. Moisture has no effect upon canisters, assuming no condensation when the canisters are taken back to the laboratory.

In the experimental program, samples recovered from the canister were desorbed onto a chromatographic column with the following parameters:

1. Column — 6 m × 0.75 mm O.D. Supelco VOCOL megabore column
2. Column temperature program — kept at 40°C for 4 min and then programmed to increase to 180°C at the rate of 4°C/min

The sample was analyzed by a GC-MS with a scan range from m/z 35–300 at 1 sec/scan in electron impact mode. Retention times and mass spectra were used to identify target compounds.

The analysis of four performance evaluation sample canisters was made using this method. The recoveries of compounds were found to be reasonably good except for gases, which had overly high percent recoveries in general, possibly due to the loss of some gases in the volatile organic standards in methanol. The authors concluded that the "method works well for the analysis of ambient air samples collected in canisters for volatile organic compounds."[20] The quantitation limit found was 2 ppb v/v or lower.

SAMPLE COLLECTION SYSTEM USING CANISTERS

Jayanty[21] discussed the design of a sample collection system using passivated SUMMA stainless steel canisters, the collection procedure, and stability data for selected organics. Also described were recent studies "using an automated cryogenic preconcentration/cryofocussing system followed by gas chromatography with a selective detector(s) for analyzing toxic organics at low ppb levels."[21]

A useful table of methods for collection of toxic organic compounds is included in Jayanty's study,[1] listing methods of collection and the major advantages and disadvantages of each. The methods of collection listed are

1. Sorbents/impingers
2. Cryogenic traps
3. Bags (Teflon, Tedlar, Mylar, etc.)
4. Glass bulbs
5. Metal canisters

It was concluded that canister-based samplers for the collection of volatile organic compounds had several advantages over the other approaches. The major advantages listed for the canisters were

1. They can be thoroughly cleaned.
2. They have good sample recovery.
3. They are rugged.
4. They can be pressurized to increase sample volume.

Also, they are not subject to sample permeation, they are not subject to photo-induced chemical effects, and they can be reused after a simple cleanup procedure. The canister is cleaned by evacuation while heating at 100°C.

The major disadvantages listed were (1) the sample size is limited, and (2) the canister is expensive.

The canister must be completely evacuated in preparation for subatmospheric sample collection. The canister is vented to the atmosphere containing toxic compounds to be sampled, the sample is caused to flow into the canister by the differential pressure between atmospheric pressure and the vacuum in the canister.

The analysis of the collected samples in the study was by automated cryogenic preconcentration followed by a chromatographic (GC) system. The GC was equipped with a flame ionization detector (FID) and an electron capture detector (ECD) connected in parallel. The parameters for the GC were

1. Column — 50 m × 0.32 mm I.D., DB-1 fused silica capillary column
2. Carrier gas — helium at 1.2 cm^3/min
3. Makeup gas — nitrogen at 26 cm^3/min
4. Detectors — an FID and an ECD in parallel
5. FID gases — hydrogen at 30 cm^3/min, air at 400 cm^3/min
6. Temperature program — 35°C for 4 min programmed at 6°C/min to 150°C
7. Cryogen — liquid nitrogen
8. Sample flow rate — 20 cm^3/min for 10 min cooled at about –160°C

Samples in a canister were found to be stable over a 10-day period, suggesting that the canisters could be used routinely for sampling, at least for the compounds tested. Sample analysis by preconcentration and GC were adequate as long as the compounds were known; otherwise gas chromatography-mass spectrometry was recommended for initial screening.

SAMPLING USING MOTOR-POWERED SYRINGES

A sampling technique based on the collection of samples in all-glass syringes with motor-driven plungers has been developed.[22]

In the technique, air to be used for the determination of solvents is drawn into the all-glass syringe (of 30-mL capacity) by drawing out the plunger at constant speed. The sample of air thus collected has a volume of 30 mL with concentrations of solvents corresponding to the mean concentrations during the sampling period. A sampling time of 1 hour was considered to be suitable; the capability to take 15-min samples should be available. The sampling device was placed either on the back or in the respiratory zone of an employee. After sampling, the syringe is sealed and removed from the sampling device.

The sample is analyzed using a gas chromatograph (GC), preferably within 4 hours of collection to avoid losses. For qualitative and quantitative analysis, normally 1 mL of the sample is injected in the GC. The GC was calibrated using aluminum laminate bags (of 5- to 10-L capacity) containing known volumes of air and known quantities of solvents.

The results of experimentation were interpreted to show that syringe pumps could be suitable for determination of concentrations of solvents in air. For conditioned syringes and analysis within 30 min after sampling, the yield could be approximately 97 to 98% if the syringes were tested for airtightness prior to sampling.

REFERENCES

1.   Des Tombe, K., D. K. Verma, L. Stewart, and E. B. Reczek. "Sampling and Analysis of Light Hydrocarbons (C_1–C_4) — A Review," *Am. Ind. Hyg. Assoc. J.* 52:136 (1991).

2.   Russo, P. J., G. R. Florky, and D. E. Agopsowicz. "Performance Evaluation of a Gasoline Vapor Sampling Method," *Am. Ind. Hyg. Assoc. J.* 48:528 (1987).

3.   Gibson, J. E., and J. S. Bus. "Current Perspectives on Gasoline (Light Hydrocarbon)-Induced Male Rat Nephropathy," *Ann. N.Y. Acad. Sci.* 534:481 (1988).

4.   Weaver, N. K. "Gasoline Toxicology — Implications for Human Health," *Ann. N.Y. Acad. Sci.* 534:441 (1988).

5.   Bertazzi, P. A., A. C. Pesatori, C. Zocchetti, and R. Lacotta. "Mortality Study of Cancer Risk among Oil Refinery Workers," *Int. Arch. Occup. Environ. Health* 61:261 (1989).

6.   Puskar, M. A., S. M. Fergon, and L. H. Hecker. "A Short-Term Determination of Hydrazoic Acid in Air," *Am. Ind. Hyg. Assoc. J.* 52:14 (1991).

7.   Choudhary, G. "Determination of 2,3-Dibromopropanol in Air," *Am. Ind. Hyg. Assoc. J.* 48:809 (1987).

8.   Hekmat, M., and R. G. Smith. "Determination of Low Molecular Weight Organic Acids Collected on Silica Gel Sampling Tubes by Using Ion-Exclusion Chromatography," *Am. Ind. Hyg. Assoc. J.* 52:332 (1991).

9.   Wood, G. O. "A Model for Adsorption Capacities of Charcoal Beds: I. Relative Humidity Effects," *Am. Ind. Hyg. Assoc. J.* 48:622 (1987).

10. Wood, G. O. "A Model for Adsorption Capacities of Charcoal Beds: II. Challenge Concentration Effects," *Am. Ind. Hyg. Assoc. J.* 48:703 (1987).

11. Yoon, Y. H., and J. H. Nelson. "Contaminant Breakthrough: A Theoretical Study of Charcoal Sampling Tubes," *Am. Ind. Hyg. Assoc. J.* 51:319 (1990).

12. Yoon, Y. H., and J. H. Nelson. "Application of Gas Adsorption Kinetics: I. A Theoretical Model for Respirator Cartridge Service Life," *Am. Ind. Hyg. Assoc. J.* 45:509 (1984).

13. Yoon, Y. H., and J. H. Nelson. "Application of Gas Adsorption Kinetics: II. A Theoretical Model for Respirator Cartridge Service Life and Its Practical Applications," *Am. Ind. Hyg. Assoc. J.* 45:517 (1984).

14. Yoon, Y. H., and J. H. Nelson. "A Theoretical Study of the Effect of Humidity on Respirator Cartridge Service Life," *Am. Ind. Hyg. Assoc. J.* 49:325 (1988).

15. Yoon, Y. H., and J. H. Nelson. "Effects of Humidity and Contaminant Concentration on Respirator Cartridge Breakthrough," *Am. Ind. Hyg. Assoc. J.* 51:202 (1990).

16. Underhill, D. W. "Calculation of the Performance of Activated Carbon at High Relative Humidities," *Am. Ind. Hyg. Assoc. J.* 48:909 (1987).

17. Werner, M. D. "The Effects of Relative Humidity on the Vapor Phase Adsorption of Trichloroethylene by Activated Charcoal," *Am. Ind. Hyg. Assoc. J.* 46:585–590 (1985).

18. Coyne, L. B., J. S. Warren, and C. S. Cerbus. "An Evaluation of a New Desorption Technique for Air Sampling Sorbent Tubes," *Am. Ind. Hyg. Assoc. J.* 48:668 (1987).

19.   Wong, J. M., N. Y. Kado, P. A. Kuzmicky, H.-S. Ning, J. E. Woodrow, D. P. H. Hsieh, and J. N. Seiber. "Determination of Volatile and Semivolatile Mutagens in Air Using Solid Adsorbents and Supercritical Fluid Extraction," *Anal. Chem.* 63:1644 (1991).
20.   Hsu, J. P., G. Miller, and V. Moran, III. "Analytical Method for Determination of Trace Organics in Gas Samples Collected by Canister," *J. Chromatogr. Sci.* 29:84 (1991).
21.   Jayanty, R. K. M. "Evaluation of Sampling and Analytical Methods for Monitoring Toxic Organics in Air," *Atmos. Environ.* 23:777 (1989).
22.   Ovrum, P. "The Sampling of Organic Solvent Vapors in Air by Motor-Powered Syringes," *Am. Ind. Hyg. Assoc. J.* 47:650 (1986).

CHAPTER 4

Respirator Cartridges and Canisters

INTRODUCTION

Respirators are used by many workers for respiratory protection in the workplace. The performance and use of respirator cartridges and canisters are reviewed in this chapter.

TESTING PROTOCOL FOR RESPIRATOR CARTRIDGES AND CANISTERS

On the premise that "the best hope for defining the overall capability of a given canister, cartridge, or pair of cartridges is to select testing conditions which will define the fundamental parameters of the carbon/vapor interactions and the influences of environmental conditions on them," a testing protocol[1] was developed which included suggested conditions and ranges of variables, and the bases for selecting them.

As prerequisites, it was assumed that [1] "(1) a testing system is set up and ready to use, (2) the test canisters, cartridges, or beds are prepared for testing, and (3) a data analysis capability exists for extracting fundamental parameters and presenting data in a useful format."

The testing protocol was limited to a single gas or vapor and to water vapor.

In developing the protocol, the objectives were to:[1]

"(1) Optimize the number of tests required to define performance for a particular vapor removal application

(2) Optimize the amount of information from testing, so that effects of environmental and use conditions on performance can be characterized

(3) Provide testing results which can be correlated by any one of a variety of mathematical models which have been or may be proposed

(4) Include enough test repetitions for at least one set of conditions that a measure of the test data reproducibility can be defined

(5) Allow for scaling of the results to another bed size, geometry, packing density, sorbent particle size, etc.

(6) Provide a visual representation of results for field use"

The canisters were to be preconditioned in such a way that they would be prepared for testing to match actual use conditions. Of particular interest was humdity preconditioning.

The protocol specifies that, for a reference set of conditions and for ranges of vapor concentration, temperature, humidity, airflow rate, and bed depth, portions of breakthrough curves be determined. A minimum number of 25 tests were considered to be required for testing to determine the performance of a selected organic vapor cartridge or canister for a selected vapor. The protocol was intended to be used to provide consistent, comparable data that would be useful for "(1) deciding on the applicability and service life of a given unit for removing a given vapor from a variety of environmental and use conditions, (2) designing new or improved canisters, cartridges, and sorbents, and (3) testing and development" of models for describing and predicting sorbent bed performance."[1]

FIELD TEST FOR DETERMINING ADSORPTIVE CAPACITY OF RESPIRATORY CARTRIDGES

The application of argon adsorption and/or desorption as a rapid, nondestructive field test of the adsorption capacity of respirator cartridges has been suggested.[2]

In experimental work, one of three adsorbents was placed in a Plexiglas container of dimensions 7.3 cm I.D. and 2.4 cm deep (the dimensions of a 100-cc Mine Safety Appliances Co. respirator cartridge). Initially, the following three adsorbents were tested:

1. Pumice, volcanic rock with little surface area
2. Calgon Carbon 12 × 40 GW, a coal-based carbon
3. Calgon Carbon ASC, a coal-based carbon impregnated with salts of chromium, copper, and silver

The experimental work was performed at room temperature.

In the first set of experiments, 3.0 L of argon was used to saturate the adsorbent with argon by flushing, followed by 1.0 L of helium. Wet-test meters were used to measure the volume of influent helium and the volume of the effluent helium plus argon gases. The difference between these two volumes was the volume of argon desorbed from the adsorbent.

In the second set of experiments, the adsorbent bed was initially saturated with helium. Then the bed was flushed slowly with argon until the volume of effluent was at least 0.3 L. The difference between the volume of the inflowing argon and the volume of the effluent gases was the volume of argon adsorbed in the second step.

The argon tests were considered to be adequate to determine nondestructively that a respirator cartridge has been filled with an adsorbent or that its adsorptive capacity is undepleted. Also, the test appeared to be simple and foolproof and did not degrade the adsorbent. The test was considered to determine only the volume of argon adsorbed.

POTENTIAL JONAS MODEL APPLICABILITY TO RESPIRATOR CARTRIDGE TESTING

The applicability of the Jonas kinetic model to the routine evaluation of the performance of organic vapor (OV) cartridges was studied experimentally.[3] In the experimental work, to resemble a packed column of varying bed length and sorbent weight, commercially available OV respirator cartridges were tested in a stacked configuration.

The kinetic adsorption capacity and gas adsorption rate of a packed charcoal adsorbent bed are described by a linear relationship between gas breakthrough time and sorbent weight in the Jonas[4-11] model. The functional equation, derived from heterogeneous gas adsorption studies, is the modified Wheeler equation. The modified Wheeler equation relates bed breakthrough time and:

1. Kinetic adsorption capacity or equilibrium adsorption capacity at an arbitrary ratio of exit to inlet concentration of gas
2. Volumetric gas flow rate
3. Weight of adsorbent
4. Bulk density of the packed bed
5. First order rate constant of adsorption
6. Inlet concentration of gas
7. Exit concentration of gas

In the laboratory setup, dried air from a main source was passed through an in-line drier. The air was varied from 70 to 110 L/min. Solvent was added to the airstream at a predetermined rate by a syringe pump to establish a known upstream concentration which was monitored by an infrared monitor. This dry airstream with challenge gas or vapor was drawn through a housing containing one to four respirator cartridges in series, providing four data points during an experimental run. An infrared monitor downstream monitored the breakthrough concentration as a function of time. Before reaching a vacuum source, the existing gas or vapors were passed through a sorbent scrubber.

Preliminary data were collected at an airflow rate of 64 L/min. Dried air and predried sorbent cartridges for acetone were used. Challenge concentrations of 530, 750, and 1060 ppm were run and 1% breakthrough times were plotted against sorbent weights. For duplicate determinations on a set of four respirator cartridges in series, the data for duplicate determinations showed excellent reproducibility. The Jonas model (modified Wheeler equation) was applied to the data.

The Jonas model for characterizing small sorbent columns at low flow rates was found to be applicable to the characterization of respirator cartridges at high flow rates. It was suggested that the Jonas model is potentially applicable for characterizing and evaluating commercially available organic vapor respirator cartridges.

ORGANIC-VAPOR MIXTURES STUDY OF RESPIRATOR CARTRIDGES

A study was made of the performance of respirator cartridges in response to single-component and binary-mixture challenges.[12] Pairs of respirator cartridges (all studied cartridges were from two lots from the same manufacturer) were tested at a flow rate of 40 L/min under carefully controlled mass flow rate, temperature, and relative humidity (RH). The flow rate was intended to represent a worst-case situation (breathing during heavy work conditions) for a respirator lifetime.

The organic vapors used in the study were

1. Methyl ethyl ketone
2. Isopropyl alcohol
3. Hexane
4. *n*-butyl acetate
5. Ethyl benzene

The mixtures used were

1. Methyl ethyl ketone and isopropyl alcohol
2. Methyl ethyl ketone and hexane
3. Ethyl benzene and *n*-butyl acetate
4. Methyl ethyl ketone and ethyl benzene
5. Isopropyl alcohol and ethyl benzene

The cartridges used were Mine Safety Appliances Co. Chemical Cartridges with a petroleum-based activated carbon adsorbent material described as having an excellent affinity for organic vapors.

The test chamber permitted the air to flow in parallel simultaneously through a pair of cartridges. The parallel exhaust airstreams were combined downstream where the combined airstream was monitored for solvent concentrations. The upstream concentrations were monitored by infrared analysis. The downstream airstream was analyzed using a gas chromatograph. The parameters for the gas chromatograph were

1. Column — 3 mm × 2 mm I.D. column
2. Column packing — glass
3. Detector — flame ionization detector

The relative humidity for the experimental work was either 50 or 85%.

For the single vapors, concentrations were 1000 ppm and then 2000 ppm. For the mixtures the concentrations were

- 1000 ppm *n*-butyl acetate and 1000 ppm ethyl benzene
- 1000 ppm isopropyl alcohol and 1000 ppm ethyl benzene
- 1000 ppm isopropyl alcohol and 1000 ppm methyl ethyl ketone
- 1500 ppm isopropyl alcohol and 500 ppm methyl ethyl ketone
- 500 ppm isopropyl alcohol and 1500 ppm methyl ethyl ketone
- 1000 ppm methyl ethyl ketone and 1000 ppm ethyl benzene
- 1000 ppm methyl ethyl ketone and 1000 ppm hexane

The observed breakthrough times for most of the mixture components were similar to those observed for each component at its high concentration. Other generalizations were that, for both moderate humidity and high humidity:

1. The smaller, lighter molecules in mixtures tended to appear sooner through the respirator cartridge than a single-component challenge would indicate
2. When two heavy molecules were mixed, both could appear sooner than anticipated from the results for single components
3. It might be possible to model organic-vapor mixture effects on respirator cartridges by using a classification scheme used for the experimental work

THEORETICAL STUDY OF EFFECT OF HUMIDITY ON RESPIRATOR CARTRIDGE SERVICE LIFE

A modified theoretical approach has been developed to investigate the effect of RH on the shelf life of respiratory cartridges.[13] The theoretical results were compared to other published experimental data for carbon tetrachloride and benzene.

A mathematical expression was developed which is applicable to asymmetrical contaminant breakthrough curves and which can be used to study the effect of RH over the entire breakthrough range of 0 to 100%. The wide range of environmental conditions for which it was possible to derive valid breakthrough curves included:

1. Respirator cartridge preconditioning humidity
2. Relative humidity of the environment in which the respirator cartridge is used
3. Flow rate through the respirator cartridge
4. Contaminant assault concentration

The validity of the theoretical model was demonstrated for carbon tetrachloride and for benzene. The theory agreed well with available experimental data for the experimental conditions in the study.

EFFECTS OF HUMIDITY AND CONTAMINANT CONCENTRATION ON RESPIRATOR CARTRIDGE BREAKTHROUGH

A model was developed to examine the effects of both test relative humidity and contaminant assault concentration on respirator cartridge breakthrough of benzene and methyl chloroform.[14] The breakthrough curve, for specified conditions of use of breakthrough cartridges, was assumed to be symmetric (rather than asymmetric as in the preceding section) and sigmoidal. Simpler theoretical expressions could be used under these conditions.

For benzene and methyl chloroform, the logarithm of breakthrough time was found to be proportional to the logarithm of concentration, with a constant of proportionality a. This, and associated expressions and theoretical parameters, permitted the generation of an entire set of contaminant breakthrough curves, each corresponding to a specific contaminant concentration for a given contaminant and a specified test humidity. This was not possible for the model in the preceding section.

The results derived from this theoretical study agreed with corresponding experimental data.

APPLICATIONS OF THE WHEELER EQUATION TO ORGANIC VAPOR RESPIRATOR CARTRIDGE BREAKTHROUGH DATA

The effect of bed weight and dry airflow rate on breakthrough curves of organic vapor respirator cartridges for a single vapor (acetone) at a single challenge concentration of 1060 ppm average, has been studied.[15] Experimental data for dried cartridges from a single manufactured lot were analyzed using three applications of the Wheeler equation. The following experimental conditions and analysis technique were included:

1. Varying sorbent bed weight
2. Varying residence time
3. Fitting the breakthrough curve

The simplified modified Wheeler equation relates bed penetration fraction to breakthrough time, with two adjustable parameters: a capacity parameter, and a rate parameter.

The cartridges used in the experimental work were dried for at least 24 hour in a vacuum oven at 100°C before testing. The acetone challenge was established by feeding the acetone at a predetermined rate to a dried airstream which passed through a cartridge cell housing containing four cartridges in series. General purpose infrared gas analyzers were used to monitor continuously the upstream and downstream acetone vapor concentration. The system was first calibrated using known concentrations of acetone in a calibrated infrared loop. Absorbance at a wavelength of 8.2 μm was monitored as a function of concentration by a digital processing oscilloscope interfaced with a digital computer. The actual experimental data were collected automatically using the digital processing oscilloscope and the computer.

Consistent values of the capacity parameter were obtained by three independent methods. The rate parameter was found to be quite dependent on the flow rate, the selected breakthrough fraction, the selected functionality of a logarithmic term, and the method of data analysis. The rate parameter was found to be approximately proportional to the square root of the airflow rate.

The modified Wheeler equation had limitations and did not characterize the system completely. The authors suggested that the modified Wheeler equation can be used by adhering to the following guidelines:

1. The method of choice for data taking and analysis is bed weight variation (by stacking two or more cartridges or canisters in series) for constant flow rate.
2. Slope, intercept, standard deviation, and linear correlation for breakthrough time vs. bed weight (or residence time) should be reported; nonlinear data fitting can be done, alternatively.
3. To allow recalculation by another model, if desired, calculated values of the capacity parameter and the rate parameter must be accompanied by descriptions of the equations and parameters used to calculate them.
4. Breakthrough times should be measured at more than one breakthrough fraction; for industrial organic vapor respirator applications, 1 and 10% were suggested.
5. In order to allow application of results to real situations (which usually differ from the laboratory test conditions), the effects of flow rate, challenge concentration, relative humidity, and temperature, in addition to bed weight variation, should be examined by varying each of these in a systematic manner.

PENETRATION OF METHYL ISOCYANATE THROUGH ORGANIC VAPOR GAS RESPIRATOR CARTRIDGES AND ACID GAS RESPIRATOR CARTRIDGES

The U.S. National Institute for Occupational Safety and Health, NIOSH, conducted research as a basis upon which to recommend protective equipment that might be used in an emergency situation where extremely high methyl isocyanate (MIC) concentrations might be encountered.[16] In particular, they studied air-purifying respirators in order to assess effectiveness against penetration of MIC vapor. MIC is "an extremely hazardous compound because of it toxicity (TLV [Threshold Limit Value] = 0.02 ppm), volatility and flammability."[16]

Penetration tests were conducted at three or four MIC challenge concentrations designed to simulate emergency escape conditions, and at three different conditions. Results were presented for two different manufacturers' organic vapor and acid gas cartridges. The Miran IA General Purpose Infrared Gas Analyzer, a single-beam variable filter infrared (IR) spectrometer capable of scanning the spectral range between 2.5 and 14.5 μm, was used as the MIC detector. The gas cell with which the spectrometer was equipped had a pathlength that varied between 0.75 and 20.25 m.

In the laboratory experimentation, house air was dried in line. The inlet airflow was controlled from 60 to 120 L/min and a syringe pump was used to inject MIC into the airstream at a predetermined rate. A known upsteam MIC concentration could be generated by adjusting the syringe pump feed rate and the inlet airflow rate. A buffer tank was added to the system to reduce fluctuations in upstream MIC concentration. The upstream MIC concentration was monitored continuously by an IR detector. A flow-temperature-humidity system was also placed in line.

The MIC-vapor-containing airstream passed through a cartridge cell housing containing a single cartridge or a pair of cartridges, depending on the design of the respirator cartridge. An IR detector downstream monitored the breakthrough concentration as a function of exposure time. The upstream and downstream IR detectors were calibrated each day before experiments were conducted. Upstream concentration varied from 280 to 1100 ppm, the downstream concentration of interest was between 0 and 50 ppm. The lower limit of detection, determined by two different methods, was 0.20 or 0.32 ppm, about a factor of ten greater than the Threshold Limit Value.

The results of the study "showed that none of the commercially available air-purifying cartridges tested provided protection against MIC breakthrough at high MIC challenge concentrations that might be expected during an emergency situation"[16] and "supported and emphasized the importance of NIOSH's recommendation that any air-purifying respirator should not be used for MIC because of the MIC's high toxicity and lack of warning properties and because no effective end of service life indicator is currently available for MIC."[16] It was recommended that "only supplied air respirators should be used, even for escape only purposes, when MIC is the suspected contaminant."[16]

CARTRIDGE RESPIRATOR PERFORMANCE FOR 1,3-BUTADIENE

The effectiveness of several types of air-purifying twin cartridges for 1,3-butadiene removal has been determined.[17] Challenge concentrations of

1,3-butadiene of 100 and 1000 ppm were developed. An infrared analyzer measured breakthrough concentration of 10 ppm.

1,3-Butadiene, which has been classified as a potential occupational carcinogen, is used in the manufacture of a variety of chemicals, rubber compounds, foams, and resins. The study was undertaken to determine the potential for air-purifying respirators to reduce exposure to 1,3-butadiene.

In the experimental work, dry clean air and deionized filtered water were supplied to a control module. In conjunction with a humidity/temperature sensor, the controller established and maintained the desired air temperature, relative humidity, and airflow rate. Airflow rates were 32.0 and 64.0 L/min. To generate the desired challenge concentrations, 1,3-butadiene flow rates ranged from 3.2 to 64.0 cm^3/min. The infrared analyzer was used to periodically check the challenge concentration. In a modeling study, three types of 12×20 mesh activated carbon were evaluated:

1. A petroleum-base carbon
2. A coal-base carbon
3. A coconut shell carbon

Three types of twin cartridges were tested to determine service life against 1,3-butadiene. Effects of flow rate, temperature, humidity and 1,3-butadiene concentration were investigated, and desorption of 1,3-butadiene from activated carbon was measured using residence time simulation. The following conclusions were drawn from the investigation:

1. The twin cartridges had reasonable adsorption capacity for 1,3-butadiene for concentrations in air of 1000 ppm or less.
2. The cartridge shelf life was reduced by both elevated temperature and elevated humidity.
3. At a given concentration, cartridge shelf life was inversely proportional to airflow rate.
4. 1,3-butadiene readily desorbed under clean air purge conditions from a saturated activated charcoal bed. The peak concentration and rate of desorption increased with increasing temperature and increasing humidity. For activated carbon beds only partially saturated with 1,3-butadiene, the onset of desorption was delayed substantially and peak concentration and the rate of desorption were reduced.

AIR-PURIFYING RESPIRATORS FOR TOLUENE DIISOCYANATE VAPORS

A disposable respirator and two brands of air-purifying organic vapor cartridges have been evaluated for protection against toluene diisocyanate (TDI) vapors.[18]

The following items summarize the experimental method:

1. Standard atmospheres of TDI concentrations of 0.2 ppm or higher were generated, with a relative humidity of 50%.
2. The cartridge or respirator was mounted so that TDI-laden air could be drawn through it at 32 L/min; the air upstream and downstream of the cartridge or respirator was monitored.
3. The breakthrough time and the useful lifetime of the cartridge or respirator were determined.

A 50-μL glass syringe containing a commercial TDI product was used to give an injection rate of 3.8 μL/hour. At TDI concentrations of 1 ppm or higher (1 to 4 ppm), the TDI atmosphere was generated by passing air over TDI in a 2000-mL round bottom flask and mixing the dry TDI-laden air with moist air to a final relative humidity of 50%. Then, 32 of 40 L/min of air coming from a mixing chamber was drawn into the cartridge or respirator by a vacuum/pressure pump. The humidity was monitored in the mixing chamber, and atmospheric pressure upstream was measured continuously. The concentration of TDI in the test atmosphere and the stability of the TDI atmosphere were quickly determined using a fast response method involving TDI tape.

Prior to drawing air through the respirator or cartridge, the TDI atmosphere was allowed to equilibrate for at least 30 min, the relative humidity of the air was adjusted to 50%, and the stability of the TDI atmosphere was checked using the TDI tape.

At a TDI concentration of 0.2 ppm, each respirator or cartridge was tested in triplicate for 40 hours; at a concentration greater than 1.5 ppm, each respirator or cartridge was tested once for 20 hours. At test times up to 40 hours, there was no breakthrough of TDI. Depending on the sampling volumes, TDI detection limits ranged from 0.0003 to 0.0009 ppm of TDI.

All the respirators or cartridges had good adsorption capacity for TDI under controlled laboratory conditions, with no facepiece leakage and no solvents or chemicals except TDI in the test atmosphere.

The results of the study showed that organic vapor respirators and cartridges could effectively absorb TDI vapors. However, there would be no sensory warning of depletion of sorbent capacity or face-seal leaks, because the odor threshold for TDI is greater than its ceiling exposure limit. "As a result, NIOSH and most of the respirator manufacturers do not recommend the use of air-purifying respirators."[18]

The authors interpret the data from this study to "indicate that NIOSH and other specialists in respiratory protection should reassess their current position against the use of air-purifying respirators in isocyanate-containing atmospheres."[18]

FIELD METHOD FOR EVALUATING SERVICE LIFE OF ORGANIC VAPOR CARTRIDGES: LABORATORY TEST RESULTS USING CARBON TETRACHLORIDE

A study was undertaken to determine whether the service life of respirator cartridges in a workplace environment can be predicted using a convenient method of workplace sampling with a carbon sorbent tube in that environment.[19]

The following questions were addressed in the study:

"(1) Is the basic premise of this study consistent with current adsorption theory and published data?
(2) Can a small respirator carbon tube (RCT) be developed that will accommodate use in field conditions by industrial hygienists?
(3) Does laboratory testing of the RCT predict how a respirator cartridge will behave?
(4) When the RCT is taken into an actual workplace, does it accurately predict the performance of a respirator cartridge?"[19]

The first three questions were answered by the study.

In a workplace, key factors that might affect the service life of respirator cartridges are

1. The presence of other contaminants that will compete for adsorption
2. The extent of the effect of water vapor on the adsorption capacity of the cartridge for a specific contaminant
3. The concentration of the contaminant of interest
4. The work (breathing) rate of the user

The method involved the use of small respirator carbon tubes (RCTs) packed with sorbent from a respirator cartridge of interest. The RCTs, unlike respirator cartridges, can be taken into the field and used with traditional high-flow air sampling pumps. The method was designed to determine the service life of respirator cartridges even in the presence of multiple contaminants in the workplace.

To determine the variability of packing of RCTs and to test the theory of bed residence time, carbon tetrachloride was used as an initial challenge agent. Carbon tetrachloride was chosen because of its use in previously published work and because it was used by NIOSH for certification testing of organic vapor cartridges. In the carbon tetrachloride testing program, one stream of air passed through a flask containing carbon tetrachloride at a constant temperature of 29 to 31°C. Carbon tetrachloride vapors passed through a condenser held at 26°C. After mixing with a second stream of conditioned air, carbon tetrachloride vapors went through a test carbon adsorbent bed (RCT, cartridge, or charcoal tube) and then into an infrared spectrophotometer.

The conclusions from the study include the following:

1. RCTs can be reproducibly packed to the same approximate packing density as that of respirator cartridges, 0.45 g/cm^3
2. RCTs were capable of predicting the performance of respirator cartridges with ±15% accuracy at the 95% confidence level

FIELD METHOD FOR EVALUATING SERVICE LIFE OF ORGANIC VAPOR CARTRIDGES: HUMIDITY EFFECTS

The effects of relative humidity (RH) and prehumidification of respirator carbon tubes (RCTs) and cartridge carbon beds on the predictive accuracy of the method discussed in the preceding section have been studied.[20] In this work, the following were examined:

1. The performance of RCTs in predicting breakthrough times of cartridges in test atmospheres of carbon tetrachloride at moderate and high relative humidity, using both "as-received" carbon from the manufacturer and prehumidified carbon
2. The effect of flow rate on the bed-residence model, over the range of bed-residence times encountered with nonpowered respirator cartridges under normal conditions of use

Organic vapor respirator cartridges with carbon packing density of 0.43 g/cm^3 ±0.01 g/cm^3 and RCTs with carbon packing density of 0.45 g/cm^3 ±0.01 g/cm^3 were used in the experimental work. Carbon tetrachloride was mixed with compressed air maintained at 21 to 24°C and 52% ±2% relative humidity for moderate humidity experiments and 92% ±1% for high humidity experiments, and passed through the carbon beds. Upstream and downstream carbon tetrachloride concentrations were monitored using infrared gas analyzers. Flow rates through the carbon beds were monitored continuously.

A stable carbon tetrachloride concentration of 1000 ppm ±60 ppm was generated at the desired relative humidity prior to exposing the test device. Exposures were continued until 100 ppm was measured downstream from the carbon bed; that is, until 10% breakthrough was attained.

RCTs using as-received carbon were tested at three different flow rates for each RH condition; cartridges were tested at two different flow rates for each RH condition. Uptakes of carbon tetrachloride and water were determined gravimetrically for each of the three or four trials for each device at a given set of exposure conditions. After prehumidification until the mass of the carbon bed increased by 25%, additional experiments were performed using RCTs and cartridges.

The authors concluded that:

1. It was demonstrated that RCTs can be used to accurately predict breakthrough times for respirator cartridges exposed to carbon tetrachloride under conditions of moderate and high humidity.
2. The fundamental adsorption behavior of the RCT is the same as that of the cartridge.

FIELD METHOD FOR EVALUATING SERVICE LIFE OF ORGANIC VAPOR CARTRIDGES: LABORATORY TESTING USING BINARY ORGANIC VAPOR MIXTURES

The effects of binary organic mixtures and intermittent testing on the predictive accuracy of the method discussed in the preceding two sections were examined.[21]

Organic vapor respirator cartridges and respirator carbon tubes (RCTs) had the same carbon packing density as in the preceding section. Also, the atmosphere generated was the same as that in the preceding section.

The predictive accuracy of the RCT method for adsorbates other than carbon tetrachloride and for binary mixtures containing carbon tetrachloride were studied in a series of laboratory tests. Pyridine, which has a stronger affinity for activated carbon than carbon tetrachloride, and normal hexane, which has a weaker affinity for activated carbon than carbon tetrachloride, and carbon tetrachloride itself were chosen for the study. Since water vapor is found in measurable quantities in most workplace environments and since the presence of water vapor can affect the adsorption of organic vapors onto carbon, a test atmosphere of approximately 50% relative humidity (RH) was selected. Testing with each of these organics and with binary mixtures of them was conducted in the continuous and the intermittent mode.

The authors concluded that:

1. "The RCT method predicted cartridge breakthrough times to an acuracy of ±17% at the 95% confidence level for *n*-hexane, pyridine, and binary mixtures of each with carbon tetrachloride in an approximately 50% RH environment."[21]
2. "Estimated kinetic adsorption capacities for each adsorbate-carbon bed system tested was reduced when a second adsorbate was introduced at the same concentration."[21]
3. "There was no evidence that *n*-hexane, carbon tetrachloride, or their mixtures exhibited migration through cartridges when exposed to a regime of testing to 50% of the breakthrough time, followed by a resting cycle of 4 days, and then a resumption of testing to breakthrough."[21]
4. "A small, statistically insignificant reduction in breakthrough time was observed when cartridges were tested intermittently with pyridine."[21]
5. "Results suggest that simple treatment of mixtures based on data from single component testing may predict incorrect breakthrough times and support the need for a field method to estimate cartridge service lives."[21]

FIELD METHOD FOR EVALUATING SERVICE LIFE OF ORGANIC VAPOR CARTRIDGES: RESULTS OF FIELD VALIDATION TRIALS

A fourth study was made of the use of respirator carbon tubes (RCTs) discussed in the preceding three sections.[22] The performance of the RCT method was examined in a workplace environment in which carbon tetrachloride concentration varied from 101 to 855 ppm and the pyridine concentration varied from 0 to 29 ppm. The range of ambient temperature was 28 to 39°C and the range of relative humidity (RH) was 19 to 49%. This was the first published study known to the authors that reported the investigation of the breakthrough time of respirator cartridges in the workplace.

Among the conclusions of the authors were

1. Cartridge breakthrough times were predicted by the RCT method to an accuracy of ±8% at the 95% confidence level; this accuracy level equalled or exceeded results of laboratory studies previously reported.
2. When ambient concentrations were standardized, a plot of residence time vs. breakthrough time yielded a coefficient of determination of 0.71.
3. "This method may prove a useful tool for industrial hygienists who wish to avoid having employees who wear APRs [air-purifying respirators] rely upon warning properties of the chemical(s) to know when safely to change their respirator cartridges."[22]

REVIEW AND COMPARISON OF ADSORPTION ISOTHERM EQUATIONS USED TO CORRELATE AND PREDICT ORGANIC VAPOR CARTRIDGE CAPACITIES

The following four adsorption equations that have been used to describe measured capacities of air-purifying cartridges were compared using experimental data:[23]

1. Freundlich
2. Langmuir
3. Dubinin/Radushkevich
4. Hacskaylo/LeVan

Air-purifying activated carbon organic vapor cartridges from three manufacturers were used in an experimental study. Five vapors were studied:

1. Ethanol, in the concentration range 275 to 2000 ppm
2. Chloroform, in the concentration range 525 to 1000 ppm
3. Carbon tetrachloride, in the concentration range 550 to 1000 ppm
4. Hexane, in the concentration range 500 to 1040 ppm
5. Acetone, in the concentration range 500 to 1745 ppm

Temperatures averaged 23°C, at 0.97 atmosphere pressure.

The authors concluded that:

1. The Langmuir, Dubinin/Radushkevich, and the Hacskaylo/LeVan adsorption equations equally satisfactorily described all of the measured concentration effects.
2. The Freundlich equation did not fit the ethanol data as well as the other three equations did.
3. The Freundlich equation has the fewest desirable characteristics.
4. The choice of equation for correlating organic vapor respirator cartridge breakthrough data may depend on which equation characteristics are most important to the user of the equation.

REFERENCES

1. Wood, G. O., and M. W. Ackley. "A Testing Protocol for Organic Vapor Respirator Canisters and Cartridges," *Am. Ind. Hyg. Assoc. J.* 50:651–654 (1989).
2. Underhill, D. W. "A Rapid Nondestructive Field Test for Determining the Adsorptive Capacity of Respiration Cartridges," *Am. Ind. Hyg. Assoc. J.* 49:235–236 (1988).
3. Moyer, E. S. "Organic Vapor (OV) Respirator Cartridge Testing—Potential Jonas Model Applicability," *Am. Ind. Hyg. Assoc. J.* 48:791–797 (1987).
4. Jonas, L. A., and J. A. Rehrmann. "Kinetics of Adsorption of Organophosphorus Vapors from Air Mixtures by Activated Carbons," *Carbon* 10:657–663 (1972).
5. Reucroft, P. J., W. H. Simpson, and L. A. Jonas. "Sorption Properties of Activated Carbon," *J. Phys. Chem.* 75:3526–3531 (1971).
6. Rehrmann, J. A., and L. A. Jonas. "Dependence of Gas Adsorption Rates on Carbon Granule Size and Linear Flow Velocity," *Carbon* 16:47–51 (1978).
7. Jonas, L. A., Y. B. Tewari, and E. B. Sansone. "Prediction of Adsorption Rate Constants of Activated Carbon for Various Vapors," *Carbon* 17:345–349 (1979).
8. Jonas, L. A., and J. A. Rehrmann. "Predictive Equations in Gas Adsorption Kinetics," *Carbon* 11:59–64 (1973).
9. Sansone, E. B., Y. B. Tewari, and L. A. Jonas. "Prediction of Removal of Vapors from Air by Adsorption on Activated Carbon," *Environ. Sci. Technol.* 13:1511–1513 (1979).
10. Bering, B. P., M. M. Dubinin, and V. V. Serpinski. "Theory of Volume Filling for Vapor Adsorption," *J. Colloid Interface Sci.* 21:378–393 (1966).
11. Dubinin, M. M. "Physical Adsorption of Gases and Vapors in Micropores," *Prog. Surf. Membr. Sci.* 9:1–70 (1975).
12. Swearengen, P. M., and S. C. Weaver. "Respirator Cartridge Study Using Organic-Vapor Mixtures," *Am. Ind. Hyg. Assoc. J.* 49:70–74 (1988).
13. Yoon, Y. H., and J. H. Nelson. "A Theoretical Study of the Effect of Humidity on Respirator Cartridge Service Life," *Am. Ind. Hyg. Assoc. J.* 49:325–332 (1988).
14. Yoon, Y. H., and J. H. Nelson. "Effects of Humidity and Contaminant Concentration on Respirator Cartridge Breakthrough," *Am. Ind. Hyg. Assoc. J.* 51:202–209 (1990).
15. Wood, G. O., and E. S. Moyer. "A Review of the Wheeler Equation and Comparison of Its Applications to Organic Vapor Respirator Cartridge Breakthrough Data," *Am. Ind. Hyg. Assoc. J.* 50:400–407 (1989).

16.   Moyer, E. S., and S. P. Berardinelli. "Penetration of Methyl Isocyanate Through Organic Vapor and Acid Gas Respirator Cartridges," *Am. Ind. Hyg. Assoc. J.* 48:315–323 (1987).

17.   Ackley, M. W. "Chemical Cartridge Respirator Performance: 1,3-Butadiene," *Am. Ind. Hyg. Assoc. J.* 48:447–453 (1987).

18.   Dharmarajan, V., R. D. Lingg, and D. R. Hackathorn. "Evaluation of Air-Purifying Respirators for Protection Against Toluene Diisocyanate Vapors," *Am. Ind. Hyg. Assoc. J.* 47:393–398 (1986).

19.   Cohen, H. J., and R. P. Garrison. "Development of a Field Method for Evaluating the Service Life of Organic Vapor Cartridges: Results of Laboratory Testing Using Carbon Tetrachloride," *Am. Ind. Hyg. Assoc. J.* 50:486–495 (1989).

20.   Cohen, H. J., E. T. Zellers, and R. P. Garrison. "Development of a Field Method for Evaluating the Service Lives of Organic Vapor Cartridges: Results of Laboratory Testing Using Carbon Tetrachloride. II: Humidity Effects," *Am. Ind. Hyg. Assoc. J.* 51:575–580 (1990).

21.   Cohen, H. J., D. E. Briggs, and R. P. Garrison. "Development of a Field Method for Evaluating the Service Lives of Organic Vapor Cartridges. III: Results of Laboratory Testing Using Binary Organic Vapor Mixtures," *Am. Ind. Hyg. Assoc. J.* 52:34–43 (1991).

22.   Cohen, H. J., S. P. Levine, and R. P. Garrison. "Development of a Field Method for Determining the Service Lives of Respirator Cartridges. IV: Results of Field Validation Trials," *Am. Ind. Hyg. Assoc. J.* 52:263–270 (1991).

23.   Wood, G. O., and E. S. Moyer. "A Review and Comparison of Adsorption Isotherm Equations Used to Correlate and Predict Organic Vapor Cartridge Capacities," *Am. Ind. Hyg. Assoc. J.* 52:235–242 (1991).

Chapter 5

Diffusion Samplers

INTRODUCTION

In this chapter, passive or diffusive sampling and uses in personal monitoring of toxic gases and vapors are reviwed.

DIFFUSION

Either of the equations:

$$J = -D\partial c/\partial x \tag{1}$$

or

$$\partial c/\partial t = D\partial c/\partial x \tag{2}$$

can be used to define the diffusion coefficient, D. Diffusion is in the x-direction, J is the diffusion flux across a unit area normal to the x-direction, $\partial c/\partial x$ is the concentration gradient of the diffusing species at a fixed time, and $\partial c/\partial t$ is the rate of change of the concentration of the diffusing species with time at a fixed distance. For J in mol cm^{-2} sec^{-1}, c in mol cm^{-3}, x in cm, and t in sec, D has units of cm^2 sec^{-1}.

Since we are generally most interested in the diffusion of organic vapors into air, the diffusion coefficient of interest is that for a binary mixture of gases A

89

(organic vapor) and B (air), D_{AB}. The binary diffusion coefficient at low to moderate pressures depends on temperature, and pressure or density.

Among the equations that have been developed to estimate values of D_{AB} is the Fuller, Schettler, and Giddings relation[1,2]

$$D_{AB} = 10^{-3} \, T^{1.75} \, [(1/M_A + 1/M_B)]^{1/2} /$$

$$P \, [(\Sigma v)_A^{1/3} + (\Sigma v)_B^{1/3}]^2$$

(3)

where T is temperature in kelvins; P is pressure in atmospheres; M_A and M_B are the molecular weights of A and B, respectively; Σv are values of the atomic diffusion volumes; and D_{AB} is in units of $cm^2 \, sec^{-1}$.

The explicit dependence of D_{AB} on T and P is shown in Equation 3. The exponent, 1.75, on T is an average value.

Example

In this example, the diffusion coefficient for water vapor in air is calculated using Equation 3. The values used for the parameters are

1. Σv for water vapor = 12.7 cm^3
2. Σv for air = 20.1 cm^3
3. M_{H_2O} = 18.0154 g/mol
4. M_{Air} = 28.964 g/mol
5. T = 299.1 K
6. P = 1 atm

$$D_{AB} = 10^{-3} \times 21511.898 \times [0.0555081 + 0.0345256]^{1/2} /$$

$$1 \times [(12.7)^{1/3} + (20.1)^{1/3}]^2$$

$$D_{AB} = 0.253 \, cm^2 \, sec^{-1}$$

The experimental value under these conditions determined by Fuller et al.[2] is 0.2580 $cm^2 \, sec^{-1}$. The calculated value, therefore, is within 2% of the experimental value.

BINARY DIFFUSION COEFFICIENTS OF VAPORS IN AIR

Lugg[3] determined experimentally the binary diffusion coefficients for 147 organic vapors and other vapors diffusing into air at 25°C and 760 mmHg. The experimental values were compared with values calculated using the prediction methods of Wilke and Lee,[4] Chen and Othmer,[5] and Hirschfelder and co-workers.[6]

Measurements of diffusion coefficients were made by the method originated by Stefan[7] and modified by others.[8-11] A stream of carrier gas (air containing less than 0.1 mg/m^3 of water vapor) was passed over the open end of a vertical, cylindrical, uniform-bore tube which contained the liquid for which the diffusion was being studied. The flow rate of the carrier gas was such that, under steady-state conditions, the vapor concentration at the open end of the cylinder was effectively zero. A concentration gradient existed down the tube to the liquid surface, where the vapor concentration corresponded effectively with the saturation vapor pressure of the liquid.[9] After a steady-state condition had been attained, the loss of liquid from the tube was measured by observing the rate of fall of the meniscus of the liquid (organic liquid in our application) using a cathetometer. The cathetometer could be read to within 0.02 mm. For liquids of low diffusion rates, measurements were made using the method of Narsimhan[12] which used a wide-diameter cell with a calibrated side arm inclined at 5° to the horizontal to reduce the time involved for measurements. The experiments were carried out at normal atmospheric pressure at a temperature of 25 ± 0.05°C. The air flow rate was 100 mL/min and the diameter of the tube was 0.517 cm.

For each compound, for each of six determinations, the position of the meniscus was noted at various times. The diffusion coefficient was calculated by the method of Desty et al.[13] from the slope of the straight line plot of the square of the diffusion path length against time in seconds. The calculated diffusion coefficients were corrected to a pressure of 760 mmHg.

The nine methods used for the calculation of diffusion coefficients of gaseous binary systems were those of:

1. Arnold[14]
2. Gilliland[15]
3. Hirschfelder, Bird, and Spotz[6]
4. Wilke and Lee[4]
5. Andrussow[16]
6. Fuller, Schettler, and Giddings[2]
7. Slattery and Bird[17]
8. Chen and Othmer[5]
9. Othmer and Chen[18]

The value of 28.967 g/mol was used for the molecular weight of air in each equation. This value is a bit high, the value of 28.964 g/mol used by Jones[19] in his development of the air density equation would be a better choice. The value 29.9 cm^3 g^{-1} mol^{-1} [20] for the molecular volume of air at its boiling point was used in the equations of Arnold; Gilliland; Hirschfelder, Bird, and Spotz; Wilke and Lee; and Andrussow.

For use in the five equations above, the molal volumes of each compound was calculated by the method of Le Bas[21] as summarized by Partington,[22] and using, wherever possible, the molecular weight, density, surface tension, and the parachor (an empirical constant relating the surface tension to the molecular volume).[23]

Atomic diffusion volumes were used in the equation of Fuller and co-workers.[2] This equation was used for those compounds for which these data were listed by the authors. The value of 20.1 cm[3] was used for the atomic diffusion volume of air.[2]

The force constants, collision diameter, and collision integral were used in the equations of Hirschfelder et al.[6] and of Wilke and Lee.[4]

Critical parameters were used in the equations of Chen and Othmer,[5] Othmer and Chen,[18] and Slattery and Bird.[17] The critical values used for air were 132.45 K, 37.2 atm, and 89.4 cm[3].[24] The equation of Othmer and Chen[18] also used the values, of the viscosity of air at the diffusion temperatures. Lugg used the value 0.01843 cP.[25]

The experimentally determined values of the diffusion coefficients are listed in Lugg.[3] The mean and standard deviation of six experimentally determined values, and the values calculated using the equations of Hirschfelder et al.,[6] Wilke and Lee,[4] and of Chen and Othmer[5] are listed for 147 organic vapors and other vapors diffusing into air.

Of the diffusion coefficients calculated using the Chen and Othmer, and Wilke and Lee equations, approximately 76 and 70%, respectively, were within ±5% of the mean experimental values. For application to high molecular weight acids, alcohols, and esters, the Hirschfelder et al.[6] equation was satisfactory. The other six equations were deemed to have limited value for the estimation of diffusivity into air of the vapors of liquids.

DEVELOPMENT OF DIFFUSION PASSIVE PERSONAL SAMPLING DEVICE

Palmes and Gunnison[26] undertook studies to develop and test "a personal monitoring device, a passive sampler based solely on molecular diffusion of gases, a principle not heretofore employed in industrial hygiene or community sampling devices." The aim of the investigation was to develop, in response to the need for personal monitoring devices in air pollution, a "novel type of device for estimating exposure to pollutant gases."

The basis of the approach was the transfer by diffusion of a gas through an orifice. For diffusion of a gas of ambient concentration, C_a, through an orifice of cross section A (in cm^2) and length L (in cm) into a chamber with concentration of C_c,

$$JA = D(A/L)(C_a - C_c) \qquad (4)$$

where D is the diffusion coefficient (cm^2/sec), J is the diffusion flux (mol cm^{-2} sec^{-1}), the units of C are mol cm^{-3}, and the units of JA are mol sec^{-1}.

The experimental setup was a chamber with an orifice at the top of the chamber through which the test gas entered the chamber. The chamber was partially filled with a substance which quantitatively collected the test gas and permitted subsequent chemical or physical estimation of the amount of gas collected. The collecting substance was a highly effective scavenger for the test gas; consequently, the concentration of the test gas in the chamber, C_c, was essentially zero and Equation 4, with $C_c = 0$, applied.

With either silica gel or concentrated sulfuric acid as the collecting substance, experiments on the transfer of water vapor were conducted. The test chambers were small vials with either glass or plastic orifices. The chambers were placed in an environment with known humidity for periods of hours to days.

Knowing the orifice dimensions, the water vapor content of the air (considered to be zero inside the chambers), the time of exposure and literature values of the diffusion coefficient of water vapor in air, the theoretical transfer of water vapor was calculated. Values of actual transfer of water vapor were determined by weight gain of the collecting substance during exposure to the diffusion of water vapor. The observed values compared well with the theoretical values, except for the smallest of the orifices.

After the feasibility of the approach had been established using water vapor, sulfur dioxide — a substance of importance as an air pollutant — was studied. Sulfur dioxide concentrations in the vicinity of 10 ppm were maintained and the exposures were for 5 hours/day. The mercuric chloride complex proposed by West and Gaeke[27] was used as the collecting substance for sulfur dioxide. The final determination of collected sulfur dioxide was by the colorimetric method described by West and Gaeke.[27] Literature values of the diffusion coefficient for sulfur dioxide in air were used in the calculations. Both wet chemical and conductimetric methods were used to monitor the maintained sulfur dioxide concentrations. Again, except for the smallest orifices, it was possible to duplicate the results obtained by wet chemical and conductimetric measurements.

The authors concluded that if a diffusion passive sampling device based on this work "were worn by a person exposed to the gas, the quantity of gas trapped in the device would be proportional to his average exposure for the period of time during which he wore the device."

VAPOR PHASE SPIKING AND THERMAL DESORPTION

Vapor phase spiking and thermal desorption of a passive sampler were studied by Gonzalez and Levine.[28] The desorption efficiency value of a solid sorbent (such as charcoal or a porous polymer) used as a collection medium, at the loading level, is used as a correction factor to account for the incomplete transfer of the adsorbed species (analyte) from the sorbent to a fluid used for desorption.

Among the most common methods used to spike the adsorbent to determine the desorption efficiency is the use of different volumes of gaseous analyte (vapor spiking). Among the methods for desorbing the analyte is thermal desorption.

Thermal desorption is much more efficient than solvent elution at low analyte loading levels, resulting in 1000-fold greater sensitivity. Ease and speed of analysis and the reuse of the sampling device are other advantages of thermal desorption.

The study of desorption efficiency and thermal desorption was performed:

"1. To compare vapor spiking to CS_2 solution liquid spiking of the analyte.
  2. To determine the viability (accuracy and precision) of thermal desorption from the coconut-based charcoal felt adsorbent used in the DuPont badge for organic vapors.
  3. To determine the desorption efficiency (DE) of a miniature passive dosimeter (MPD) for organic vapors."[28]

EVALUATION OF SAMPLING RATE CALCULATIONS BASED ON ESTIMATED DIFFUSION COEFFICIENTS

Methods of calculating sampling rates of commercially available diffusive samplers for organic vapors based on estimated diffusion coefficients have been studied by Feigley and Lee.[29] The Brokaw[30] and Fuller and co-workers[2] methods were used to calculate estimated diffusion coefficients. The compounds of interest were all those for which experimentally measured sampling rates were available for 3M and DuPont samplers, if sufficient data could be obtained. For each of the two methods and each of the two sampler brands, least-square, simple linear regression equations for calculating sampling rates based on estimated diffusion coefficients were computed separately. Also, a regression equation for predicting the sampling rates of the MSA Vaporgard samplers was calculated. Only six empirically determined rates were available for these samplers and only the Fuller et al.[2] estimates of diffusion coefficients were used.

Regressions were run for 3M and DuPont samplers using a model with separate coefficients for each of three chemical groupings:

1. Alcohols, aliphatics, Cellosolves, esters, and ketones
2. Aromatic and cyclic compounds
3. Halogenated compounds

Also, the diffusion volume increment for bromine was estimated from the measured diffusion coefficients for seven brominated organic compounds.

Measured sampling rates were plotted against estimated diffusion coefficients for the DuPont sampler and for the 3M sampler, along with lines of regression.

The two methods used to estimate diffusion coefficients were equally effective in fitting sampling rate data for both the DuPont and the 3M samplers. However, the accuracy of estimating sampling rates for the DuPont samplers "had much greater variability than the accuracy of estimating rates for the 3M samplers."[29]

Sampling rates calculated from the diffusion coefficient multiplied by the physical area-to-length (A/L) ratio of the sampler averaged 27 to 56% higher than the actual measured rates for the DuPont and 3M samplers. Theoretically determined sampling rates would be higher than measured sampling rates for the MSA sampler also.

Estimation of sampling rates using the regression equations was considered to be preferable to several other approaches using diffusion coefficient data. The use of the Fuller et al.[2] method was recommended and it was extended for use with organic vapors containing bromine.

EFFECT OF ATMOSPHERIC PRESSURE ON COLLECTION EFFICIENCY OF A DIFFUSIONAL SAMPLER

In the use of diffusional samplers it had been assumed that there was no change in sampling efficiency due to changes in pressure.[31] Lindenboom and Palmes[31] experimentally studied the effect of atmospheric pressure on the collection efficiency of a simple diffusional sampler in the pressure range of 1 atm to less than 0.1 atm.

The samplers were nitrogen dioxide samplers.[32] Two or more chambers connected in series so that the same airstream flowed through all of the chambers constituted the exposure system for the experiments. Each of the chambers was fitted with a diaphragm-type vacuum gage with readings from 1 to 0 atm.

The approach taken was to expose samplers in one chamber (flask) to a stream of air containing nitrogen dioxide and then pass the effluent from that flask to a second flask. The pressure in the second flask was maintained at 1 to approximately 0.1 atm.

The sampler efficiency was found to be dependent on atmospheric pressure— decreasing as pressure decreases. For a diffusion path of constant diameter, the shorter the pathlength the greater the effect of atmospheric pressure on collection efficiency. Since the loss in efficiency was found to be less than 7% at an altitude of 18,000 ft, a correction for pressure would not be necessary near the surface of the earth.

Other results are

1. There was decreased collection efficiency although relative concentrations of nitrogen dioxide air remained constant.
2. Reduced sampler collection at reduced atmospheric pressure was not due to revolatilization of nitrogen dioxide.
3. Nitrogen dioxide concentration, flow rate, temperature, and wall loss were independent of pressure.
4. As the length of the sampler was reduced, collection efficiency became lower.

COLLECTION DEVICE FOR SEPARATING AIRBORNE VAPOR AND PARTICULATES

A collection device, appropriate for use in personal sampling and used to distinguish between vapor and aerosol has been developed and evaluated.[33] The device is a two-stage sampler which separates vapor and particles into two identifiable fractions, using the differential diffusion rate of vapor and particles in air.

The first stage is a glass tube with sorbent-coated walls. The vapor phase diffuses to the sorbent and is trapped on it. Particulates, which have lower diffusion coefficients, pass through the tube.

In the second stage, the particulates passing from the tube were collected on a standard 37-mm filter in a cassette.

The collector was tested, with testing focused on the first, vapor collection, stage. A material in the vapor phase only (aniline, for the aromatic amine sampler) was used for testing. Iron oxide fume was used in separate experiments to determine particulate passing efficiency. The test atmospheres were sampled with vapor collector tubes.

Airborne material was separated into vapor and aerosol fractions by appropriate choice of sampling flow rate, tube diameter, and tube length. The sampler was found to be practical for monitoring worker exposure and for inhalation studies for which monitoring of the mode of exposure and the concentration were desirable.

RELIABILITY OF BACKUP SYSTEMS IN DIFFUSIVE SORBENT SAMPLERS

The ability of passive monitor backup systems to provide users with greater assurance of sampling reliability were investigated by Guild et al.[34] The backup systems in use by current manufacturers of passive monitors including 3M, Pro-Tek, and SKC, were studied.

Two basic backup system types in use were (1) a secondary adsorbing layer behind the main adsorbing layer, and (2) a dual-sample backup system.[35]

The secondary adsorbing layer was intended to adsorb sample not adsorbed in the primary adsorbing layer. Two separate monitors were provided in the dual-sample backup system.

The experimental apparatus was designed to expose a passive sampler to a known concentration of a chemical under controlled:

1. Concentration
2. Temperature
3. Humidity
4. Wind velocity
5. Time

6. Multicomponent interferences
7. Monitor orientation

The three major components of the experimental apparatus were (1) an air-conditioning chamber for the airflow stream, (2) a chemical injection system, and (3) an exposure chamber.

Methylene chloride was used in the study as a representative low-molecular-weight compound that was difficult to sample by passive monitors. After exposure to the methylene chloride, all samples were desorbed with 2.0 mL of carbon disulfide and analyzed by gas chromatography. The operating parameters for the gas chromatograph were

1. Column — 10% Carbowax 20M on 80/100 mesh Chromosorb W-AW
2. Column temperature — 70°C
3. Injector temperature — 125°C
4. Detector temperature — 125°C
5. Detector — Flame inoization detector

The sampling accuracy of passive monitors for the collection of methylene chloride was evaluated in a series of experiments. The results for the experiments indicated that the use of passive monitors with single adsorbent layers for the collection of methylene chloride can result in gross sampling errors. The effects of reverse diffusion, which are enhanced by the presence of interfering compounds, were considered to be the cause of errors in sampling. The usefulness of backup systems as a means of improving sampling reliabilty was found to be severely limited.

PERSONAL MONITORING USING A DIFFUSIVE SAMPLER AND SHORT-TERM STAIN TUBE

The suitability of a technique involving the collection of an analyte on a diffusive sampler followed by thermal desorption onto a short-term stain tube was demonstrated for three chlorinated hydrocarbons:[36]

1. 1,1,1-Trichloroethane
2. Trichloroethylene
3. Tetrachloroethylene

The diffusive samplers were coupled with a carrier gas supply to a gas chromatograph. A known mass of analyte was injected directly onto the sampler and then swept through the chromatograph column for detection by flame ionization. This technique was repeated for each analyte and the adsorbers over the range of sampler temperatures of 50 to 250°C.

The stain tubes used were the appropriate Draeger tubes.

To obtain a measure of recovery efficiency of the samplers, the chromatographic peak areas were those obtained for the same mass of analyte injected directly onto the column.

To evaluate the dosimeter as a complete sampling and analysis system, the diffusive sampler was exposed in a series of experiments to a range of analyte concentrations over various time periods. Subsequent analysis was carried out by desorption onto a stain tube. The three chlorinated hydrocarbons were studied for exposure times between 6 and 16 hours at analyte concentrations between about 10 and 300% of the respective 8-hour exposure limits. The adsorbents (and desorption temperatures) used were Chromosorb 102 and 106 (150°C) for 1,1,1-trichloroethane, Chromosorb 102 and 106 (200°C) for trichloroethylene, and Chromosorb 102 and Tenax GC (200°C) for tetrachloroethylene.

The authors of the description of the study concluded that "when the criteria are met and the limitations are considered, the technique can form the basis of a simple, versatile and low cost method of personal monitoring."

COMPARISON OF ORGANIC VAPOR MONITORS WITH CHARCOAL TUBES FOR SAMPLING OF BENZENE IN AIR

A laboratory comparison of the 3M 3500 organic vapor monitor with charcoal tubes was made in both short-term and full-shift scenarios.[37] Because of the general interest in benzene in both the chemical and refining industries, it was chosen as the analyte. This choice was also appropriate because of the presence of benzene in environmental tobacco smoke. Pristas,[37] the author of the account of the work, attempted to address those items that were considered to be major concerns of industrial hygienists.

Dynamic sample atmospheres were generated by (1) dilution of a primary standard benzene-in-air gas cylinder mixture, or (2) introduction of neat benzene into a flowing airstream by the use of a syringe pump or a peristaltic device.

In long-term time-weighted average (TWA) studies, test atmospheres were verified by whole gas injections into a gas chromatograph against gas standards. In short-term exposure limit (STEL) studies, chamber concentrations were verified by a combination of whole gas injections on a gas chromatograph with flame ionization and the use of a hydrocarbon analyzer to monitor the sampling chamber.

The possible effect or relative humidity (RH) on passive badge sampling rates for benzene was evaluated using two humidity levels: 4% RH at 37°C, and 60% RH at 37°C. Using six different concentration levels (0.05 to 8.4 ppm), the effects of concentration were evaluated through side-to-side monitoring of the exposure chamber. Reverse diffusion (the loss of analyte from a previously exposed passive badge during periods of low or zero analyte concentration after the exposure to the analyte) was evaluated. Monitor sampling effectiveness in the presence of competing solvents (toluene, ethyl benzene, and three xylene isomers) was evaluated. At 0, 14, and 28 days, storage stability at room temperature was checked.

Several different studies were conducted to "evaluate the effectiveness of the passive badges relative to charcoal tubes for measuring the typical types of exposure profiles that might be encountered during an actual work shift or practice":

1. Steady-state exposures
2. Spikes of 5-min and 2-min duration in conjunction with steady-state exposure
3. Multiple short-duration spikes of less than 15-sec duration
4. Varying concentration spike clusters
5. Evaluation of the collection of benzene in the presence of gasoline
6. Evaluation of the ability of the passive badges to monitor fluctuating concentrations
7. Evaluation of the precision and accuracy of the passive badges and charcoal tubes under a short-duration (less than 15-sec) single intense exposure that may be instantaneous

Under both high and low humidity, experimental sampling rates agreed with the manufacuter's published value. Side-to-side sampling of six benzene concentrations, from 50 ppb to 8.4 ppm, gave comparable results for the monitors and the charcoal tubes. There was no evidence of reverse diffusion. In the presence of competing solvents, the results were comparable to those obtained in benzene-only challenges. There was no loss in storage over a 28-day period at room temperature.

The 3M badges had a slightly higher bias than the charcoal tubes for exposures of a 15-min duration to a 0.53-ppm benzene challenge. They were also less precise than the charcoal tubes. The benzene exposure generated by the spike and the low-level benzene background were effectively integrated by both the 3M badges and the charcoal tubes. There was good correlation of the data for both benzene and gasoline in the benzene-enriched gasoline studies. The 3M badges had a "slightly" higher negative bias, and were less precise, than that of the charcoal tubes for the benzene-only single-spike evaluation. For the multicomponent challenge, the 3M badges had the lower biases and the charcoal tubes were the more precise. Both types of sensors had negative biases and good precision for the benzene-only spike cluster evaluation. Both types of sensors had negative biases for the multicomponent cluster challenges, but the charcoal tube was the more precise. In long-term studies, the 3M badges were generally the more precise. In short-term studies, the charcoal tubes were the more precise.

The 3M badges gave precise and accurate results that are well within the ±25% accuracy acceptability criterion of the National Institute for Occupational Safety and Health (NIOSH).

For the evaluation of benzene in air, on the basis of the experimental data, the 3M 3500 passive badge can be expected to yield comparable results to those of a charcoal tube.

DESIGN OF DIFFUSIVE SAMPLER FOR PERSONAL MONITORING OF TOLUENE DIISOCYANATE EXPOSURE

A passive dosimeter has been designed for personal monitoring of toluene diisocyanate (TDI).[38] The required characteristics were that it be simple and inexpensive to build, and exhibit reproducible performance with precision and accuracy comparable to recommended active sampling techniques.

The design was governed by:

1. Choice of collection media
2. Collection of regulating membrane
3. Elimination of problems in sampling including:
   a. Face velocity effects
   b. Differences in isomer response
   c. Residual TDI in the membrane

The dosimeter was constructed from a three-piece, 37-mm aerosol cassette. The collection medium was 0.5% sulfuric acid. Microporous Teflon diffusion membranes were used, and a second microporous membrane was used as a windscreen.

The TDI collected was determined as toluenediamine by one of two techniques: (1) a spectrophotofluorometric method using the fluorescent reagent fluorescamine, and (2) a spectrophotometric method based on the Marcali technique.[39]

The mass transfer rate of TDI was approximately $0.0152\ \mu g/[(ppb) \cdot (hour)]$, equivalent for the 2,4- and 2,6- isomers of TDI. The authors concluded that the passive dosimeter "shows promise for the monitoring of occupational exposure to TDI in epidemiologic surveys and in routine assessment of compliance efforts."

LABORATORY AND FIELD EVALUATION OF THE TDI DOSIMETER

Rando et al.[40] described a laboratory and field evaluation of the toluene diisocyanate (TDI) dosimeter described in the section above. The evaluation included an assessment of:

1. Accuracy and precision
2. The effects of ambient conditions:
   a. Temperature
   b. Humidity
   c. Atmospheric pressure
3. The effects of chemical interferences
4. A field test at a TDI manufacturing facility

A study of dose-response calibration was made using three of the dosimeters. In this study, the dosimeters were exposed to test atmospheres of 2,4- and 2,6-TDI in a bench-top exposure chamber constructed from a 2-L water-jacketed reaction kettle. The flow, maintained at constant temperature and constant humidity, was at a rate of 45 L/min. The estimated face velocity of the airstream flowing past the dosimeters was 0.18 m/sec (35 ft/min). The TDI was generated using permeation cells and its concentration in the chamber was monitored continuously.

The temperature and absolute humidity of the test atmosphere were maintained at 25.0°C and 12 mmHg (50% RH), respectively. The target concentration was 20 ppb for each isomer of TDI and the exposure period for the dosimeters was 4 hours followed by a 0.5-hour sitting time before emptying. The Mercali method,[39] modified for the dosimeters, was used to analyze the samples.

After the preliminary dose-response calibration, the calibration was expanded over a wide range of exposure doses (from 4.1 to 94.0 ppb), exposure periods of from 1.0 to 6.0 hours, and an equilibration period of 30 min prior to removal of the absorbing medium and analysis. The evaluation was performed at 25°C and 50% RH.

The response of the dosimeter was found to be linear up to an exposure dose of at 300 ppb/hour. The expected precision of the response of the dosimeter was well below (better than) that described in the NIOSH sampling strategy manual.[41]

The influence of temperature on the performance of the dosimeter over the temperature range of 6 to 35°C is expressed by a temperature correction factor of −1.0%/°C. There was significant correlation of the amount of TDI collected with atmospheric pressure, after accounting for exposure dose. No dependence of dosimeter performance on humidity level, for either isomer, was found. Negative interference by ambient nitrogen dioxide was eliminated by addition of sulfamic acid to the collecting medium. N-ethylmorpholine was found to be a positive interferent, resulting in indicated levels of TDI about 25% higher than true levels.

The dosimeter was field tested at an industrial chemical plant manufacturing toluene diisocyanate by phosgenation of toluenediamine. Samples were collected with dosimeters and impingers, side by side, for Mercali analysis. When the TDI was in the vapor phase only, there was good correlation with the reference method. However, the device was found to be susceptible to a large positive bias if the samplings were done in an atmosphere containing aerosolized TDI.

REVIEWS OF PASSIVE DOSIMETRY, DIFFUSIVE SAMPLING

Two reviews, published in 1982 by Rose and Perkins[42] and in 1987 by Harper and Purnell,[43] are useful in tying together much of the material in the previous sections of this chapter and giving an overall view of diffusive sampling.

Rose and Perkins Review

The personal passive dosimeter has been called "personal" because it can be worn on the person of a worker close to the breathing zone, and it has been called "passive" because the air is not moved over the collector by a pump. The dosimeter is appealing because of its simplicity and the absence of a need for elaborate calibration procedures.

Five factors affecting the measurement of the concentration of a substance in ambient air using a diffusional monitor are

1. The length of the diffusion path
2. The cross-sectional area of the diffusion path
3. The total mass of the substance collected by the monitor
4. The length of time the monitor is exposed to the contaminated atmosphere
5. The diffusion coefficient

For permeation dosimeters that rely on the permeation of a contaminant through a membrane, the factors affecting the measurement of the concentration of a substance in ambient air are

1. The total mass of the substance collected by the monitor
2. The length of time the monitor is exposed to the contaminated atmosphere
3. The permeation constant

The measurements or determinations of the eight factors listed above are obvious sources of error for the two types of monitors. Another source of error is the presence of interferences. The face velocity, the velocity of the air external to the monitor, is a final source of error. For passive collectors, the determination of the diffusion coefficient and the face velocity are unique sources of error affecting the calculation of concentration of a substance in ambient air. Another factor of importance is the relative humidity of the sampled air.

The precisions of passive dosimeters are essentially equivalent to those of conventional techniques. The additional 5% error associated with mechanical pumps, in many cases, makes the passive dosimeters even more attractive. In general, available data indicate that passive dosimetry is an acceptable method for monitoring gases and vapors.

Harper and Purnell Review

The worst interfering substance in diffusive sampling is water vapor, which is present in nearly all occupational sampling situations.

"Diffusive samplers can be used over a wide range of exposure dosages (i.e., exposure × time, measured in ppm hours or ppm minutes). The critical limitations are sorbent capacity and uptake rate in the case of high dosages and the response time and sensitivity in the case of low dosages."[43] Exposures over a full work shift (8 hours, for example) can be integrated by diffusive samplers. For use in short-term measurements, any sampler must have a fast response time and the ability to accurately assess transient concentrations.

Possible sources of errors in the use of activated charcoal in diffusive sampling include:

1. Sorbent capacity
2. Adsorbate competition
3. Storage loss
4. Humidity

In addition to the other advantages already named, diffusive samplers can furnish an immediate assessment of exposure.

The authors[43] conclude that "diffusive sampling systems have enormous future potential in occupational hygiene monitoring."

REFERENCES

1. Fuller, E. N., and J. C. Giddings. *J. Gas. Chromatog.* 3:222 (1965).
2. Fuller, E. N., P. D. Schettler, and J. C. Giddings. *Ind. Eng. Chem.* 58:19 (1966).
3. Lugg, G. A. "Diffussion Coefficients of Some Organic and Other Vapors in Air," *Anal. Chem.* 40:1072–1077 (1968).
4. Wilke, C. R., and C. Y. Lee. *Ind. Eng. Chem.* 47:1253 (1955).
5. Chen, N. H., and D. F. Othmer. *J. Chem. Eng. Data* 7:37 (1962).
6. Hirschfelder, J. O., R. B. Bird, and E. L. Spotz. *Trans. Am. Soc. Mech. Eng.* 71:921 (1949).
7. Stefan, J. *Ann. Physik* 41:723 (1890).
8. Altshuller, A. P., and I. R. Cohen. *Anal. Chem.* 32:802 (1960).
9. Hudson, G. H., J. C. McCoubrey, and A. R. Ubbelohde. *Trans. Faraday Soc.* 56:1144 (1960).
10. Lee, C. Y., and C. R. Wilke. *Ind. Eng. Chem.* 46:2381 (1954).
11. Richardson, J. F. *Chem. Eng. Sci.* 10:234 (1959).
12. Narsimhan, G. *Trans. Indian Inst. Chem. Eng.* 8:73 (1955–56).
13. Desty, D. H., C. J. Geach, and A. Goldup. *Gas Chromatography, 1960,* R. P. W. Scott, Ed., (London: Butterworths, 1960).
14. Arnold, J. H. *Ind. Eng. Chem.* 22:1091 (1930).
15. Gilliland, E. R. *Ind. Eng. Chem.* 26:681 (1934).
16. Andrussow, L. *Z. Elektrochem.* 54:566 (1950).
17. Slattery, J. C., and R. B. Bird. *A. I. Ch.E. J.* 4:137 (1958).
18. Othmer, D. F., and H. T. Chen. *Ind. Eng. Chem. Process Design Develop.* 1:249 (1962).
19. Jones, F. E. "The Air Density Equation and the Transfer of the Mass Unit," *J. Res. Natl. Bur. Stand. (U.S.)* 83:419 (1978).
20. Treybal, R. E. *Mass Transfer Operations* (New York: McGraw-Hill, 1955) p. 27.
21. Le Bas, G. *The Molecular Volumes of Liquid Chemical Compounds* (New York: Longman, Green, 1915).
22. Partington, J. R. *An Advanced Treatise on Physical Chemistry,* Vol. II (New York: Longman, Green, 1951) p. 22.

23. Sugden, S. *The Parachor & Valency* (London: Routledge, 1930) pp. 30–32.

24. Reid, R. C., and T. K. Sherwood. *The Properties of Gases and Liquids. Their Estimation and Correlation,* 2nd ed. (New York: McGraw-Hill, 1966).

25. *Tables of Thermal Properties of Gases,* National Bureau of Standards Circular 564 (Washington: U. S. Government Printing Office, 1955) pp. 10–11, 69.

26. Palmes, E. D., and A. F. Gunnison. "Personal Monitoring Device for Gaseous Contaminants," *Am. Ind. Hyg. Assoc. J.* 34:78–81 (1973).

27. West, P. W., and G. C. Gaeke. "Fixation of Sulfur Dioxide as Disulfitomercurate (II) and Subsequent Colorimetric Estimation," *Anal. Chem.* 28:1816 (1956).

28. Gonzalez, J. A., and S. P. Levine. "Vapor Phase Spiking and Thermal Desorption of a Passive Sampler," *Am. Ind. Hyg. Assoc. J.* 48:739–744 (1987).

29. Feigley, C. E., and B. M. Lee. "Determination of Sampling Rates of Passive Samplers for Organic Vapors Based on Estimated Diffusion Coefficients," *Am. Ind. Hyg. Assoc. J.* 49:266–269 (1988).

30. Reid, R. C., J. M. Prausnitz, and T. K. Sherwood. *The Properties of Gases and Liquids* (New York: McGraw-Hill, 1977).

31. Lindenboom, R. H., and E. D. Palmes. "Effect of Reduced Atmospheric Pressure on a Diffusional Sampler," *Am. Ind. Hyg. Assoc. J.* 44:105–108 (1983).

32. Palmes, E. D., A. F. Gunnison, J. DiMattio, and C. Tomczyk. "Personal Sampler for NO$_2$," *Am. Ind. Hyg. Assoc. J.* 37:570–577 (1976).

33. Gunderson, E. C., and C. C. Anderson. "Collection Device for Separating Airborne Vapor and Particles," *Am. Ind. Hyg. Assoc. J.* 48:634–638 (1987).

34. Guild, L. V., D. F. Dietrich, and G. Moore. "Assessment of the Reliability of Backup Systems in Diffusive Sorbent Samplers," *Am. Ind. Hyg. Assoc. J.* 52:198 (1991).

35. Moore, G., S. Steinle, and H. Lefebre. "Theory and Practice in the Development of a Multisorbent Passive Dosimeter System," *Am. Ind. Hyg. Assoc. J.* 45:145 (1984).

36. Gentry, S. J., and P. T. Walsh. "Eight-Hour TWA Personal Monitoring Using a Diffusive Sampler and Short-Term Stain Tube," *Am. Ind. Hyg. Assoc. J.* 48:287–292 (1987).

37. Pristas, R. "Benzene in Air — Organic Vapor Monitors versus Charcoal Tubes," *Am. Ind. Hyg. Assoc. J.* 52:297–304 (1991).

38. Rando, R. J., Y. Y. Hammad, and S.-N. Chang. "A Diffusive Sampler for Personal Monitoring of Toluene Diisocyanate (TDI) Exposure. Part I: Design of the Dosimeter," *Am. Ind. Hyg. Assoc. J.* 50:1–7 (1989).

39. Mercali, K. "Microdetermination of Toluenediisocyanate in Atmosphere," *Anal. Chem.* 29:552–558 (1957).

40. Rando, R. J., Y. Y. Hammad, and S.-N. Chang. "A Diffusive Sampler for Personal Monitoring of Toluene Diisocyanate (TDI) Exposure. Part II: Laboratory and Field Evaluation of the Dosimeter," *Am. Ind. Hyg. Assoc. J.* 50:8–14 (1989).

41. Leidel, N. A., K. A. Busch, and J. R. Lynch. *Occupational Exposure Sampling Strategy Manual,* NIOSH Pub. No. 77–173, (Cincinnati, Ohio: National Institute for Occupational Safety and Health, 1977) p. 78.

42. Rose, V. E., and J. L. Perkins. "Passive Dosimetry — State of the Art Review," *Am. Ind. Hyg. Assoc. J.* 43:605–621 (1982).

43. Harper, M., and C. J. Purnell. "Diffusive Sampling — A Review," *Am. Ind. Hyg. Assoc. J.* 48:214–218 (1987).

CHAPTER 6

Environmental Tobacco Smoke

INTRODUCTION

One of the most prevalent sources of pollution of air of particular concern indoors in the workplace, in the home, in restaurants, in public buildings, in buses, on trains, on airplanes, etc. is smoke from cigarettes, cigars, and pipes. Smoking has now been banned on domestic flights in the U.S. and in some other areas. By far the most prevalent of smoking pollutants are those from cigarette smoking. In this chapter, we shall concentrate on organic gases in cigarette smoke.

ENVIRONMENTAL TOBACCO SMOKE (ETS)

Tobacco smoke in the environment is called environmental tobacco smoke (ETS), to which smokers and nonsmokers (also referred to as involuntary smokers and passive smokers) are exposed.

ETS originates at the lighted tip of the cigarette, and exposure is greatest in the proximity of the smoker.[1] More than 3800 compounds have been identified in cigarette smoke.[2] ETS exists in two phases: the vapor phase and the particulate phase. It can be a substantial contributor to the level of indoor pollution concentrations of benzene, acrolein, N-nitrosamine, carbon monoxide, and respirable particles. It is the only source of nicotine and some N-nitrosamine compounds in the general environment.[2] ETS is diluted by the larger volume of air with which it mixes and it ages prior to inhalation.[1] ETS is the combination of the smoke exhaled by the smoker, the smoke emitted by the burning end of a cigarette between puffs, smoke that escapes from the burning end during puff-drawing, and gases that diffuse through the cigarette paper during smoking.[1,2]

Mainstream Smoke

Mainstream smoke (MS) is the complex mixture that exits from the mouth-piece of a burning cigarette and is drawn through the tobacco into the smoker's mouth when a puff is inhaled by the smoker.

Sidestream Smoke

Sidestream smoke (SS) is the smoke emitted by the burning tobacco between puffs.[1] Since the temperature of combustion during sidestream smoke formation is lower than that for mainstream smoke formation, greater quantities of organic constituents of smoke are generated. For example, SS contains greater amounts of nicotine, benzene, carbon monoxide, N-nitrosamine, ammonia, 2-naphthy-lamine, 4-aminobiphenyl, benz[a]anthracene, and benzo-pyrine per milligram of tobacco burned.[1] SS contains more free nicotine in the vapor phase than does mainstream smoke.

Human Exposure to Environmental Tobacco Smoke

Human exposure to ETS can be estimated using approaches similar to those used for the airborne pollutants. However, no single compound definitively characterizes an individual's exposure to ETS.[1]

There is currently no direct measure of the dose of ETS absorbed in a population under study.[2] Exposures can, however, be assessed by air monitoring and other means.[2] The use of air monitoring (personal or indoor space) is handicapped by the lack of a clear definition of the physicochemical nature of ETS and the identification of the individual, or target, constituents of ETS.[2] Surrogate constituents have been measured as indicators of exposure to environ-mental tobacco smoke in both indoor and space monitoring.[2] Both active and passive monitors can be used to measure an individual's exposure to an air contaminant at the breathing zone.

Proxies, Surrogates, Tracers, and Markers

A number of proxy, or surrogate, constituents have been measured in a number of studies as indicators of ETS in both personal and indoor space monitoring. They have also been referred to as markers or tracers. No single marker has quantified accurately the exposure to each of the constituents of smoke over the wide range of environmental settings in which smoking occurs.[1]

Markers should be chosen both because of their accuracy in estimating exposure and because of their relevance for the health outcome of interest.[1] An ideal marker should be unique (or nearly unique) to tobacco smoke, should be a constituent of tobacco that is present in sufficient quantity that it can be measured even at low levels of ETS, and should stand in constant ratio across brands of cigarettes to other tobacco smoke constituents or contaminants of interest.[2]

Certain gases have been measured as possible indicators of ETS exposure, including formaldehyde and acrolein, and aromatic compounds such as benzene, toluene, xylene, and styrene.[1]

Carbon monoxide (CO) has been measured extensively, both in indoor chamber studies and in occupied public and nonindustrial occupational indoor spaces, to represent levels of ETS.[2] Under steady-state conditions in chamber studies, where outdoor CO levels are known, and the tobacco brands and smoking protocols are constant, CO can be a reasonably reproducible indicator of ETS exposure.[2]

Nicotine appears to be a promising tracer for ETS because of its specificity for tobacco and its presence in relatively high concentrations in tobacco smoke.[1] At a practical level, the technology for measuring nicotine levels is available and accurate. Nicotine volatilizes during dilution of sidestream smoke, so that it occurs almost exclusively in the vapor phase.[1] Almost all of the nicotine shifts from the particulate phase in MS and fresh SS to the vapor phase in ETS.[2] Tobacco is the only source of nicotine, so the *Nicotiana* alkaloid is a specific indicator for tobacco smoke pollution.[2]

EXPOSURE TO ETS

Use of Diffusion Denuder Samplers

Diffusion denuder samplers were used by Eatough et al.[3] to collect gas-phase acids and bases separately from particle-phase acids and bases present in environmental tobacco smoke.

ETS was sampled from a 10-m^3 unventilated environmental chamber in the initial experiment; a 30-m^3 unventilated chamber was constructed and used for all subsequent chamber experiments.

The environmental chambers consisted of Teflon bags with Teflon sampling manifolds at the bottom of the bag. Before beginning an experiment, the air quality in the chamber was checked to assure that the background was negligible compared to the ETS to be produced. After approximately 30 min of background data collection, a single reference cigarette in the 10-m^3 chamber, or 1 to 4 cigarettes in the 30-m^3-chamber, were lit electrically and burned with either a standard puffing cycle or a 2-sec 35-cm^3 puff every minute, or a 10-sec initial puff followed by 7 2-sec puffs at 1-min intervals.

The mainstream smoke was withdrawn from the chamber, the sidestream smoke was allowed to mix freely in the chamber. The cigarette(s) was(were) extinguished with water at the end of the 5- to 7-min combustion period. The chamber air was mixed by a Teflon-coated stirrer at 50 rpm for either 1 min after completion of the smoking cycle or for the length of the 8-min smoking cycle. The chamber air was then allowed to equilibrate, without mixing, for about 15 min. A cylindrical and/or annular diffusion denuder was used to sample the atmosphere in the environmental chamber for 1 to 4 hours. In some experiments, a total hydrocarbon analyzer was used to monitor changes in gas phase organic compounds.

Sections of glass tubing 10 cm in length with an inside diameter of 0.56 ± 0.01 cm were used to construct cylindrical diffusion denuders. The inside wall was coated with 0.8 M benzenesulfonic acid solution, drained, and then dried with a nitrogen stream, for the collection of gas-phase organic bases.

The annular space of annular diffusion denuders was coated with 0.8 M benzenesulfonic acid solution for the collection of nicotine and other gas-phase organic bases. The annular space was 0.13 cm wide; the annular denuder sections were 20 cm long with an inside diameter of 3.18 cm.

The denuder systems collected nicotine from prepared gaseous samples. Gas chromatography (GC) and ion chromatography (IC) were used to determine nicotine in the collected samples. Two different capillary columns, a flame ion detector (FID) or nitrogen-phosphorus detector (NPD), and an integrator were used for GC determinations of nicotine.

Initial identification of compounds in the benzenesulfonic acid denuder extracts was made using gas chromatography-mass spectroscopy (GC-MS). An MPIC column and an ultraviolet (254 nm) detector were used in ion chromatographic determination of nicotine.

There were 16 gas-phase organic bases detected, and 14 were identified in the extracts, in addition to nicotine. Quantitative results for seven gas-phase organic nitrogen compounds were obtained by GC-FID. An annular denuder/filter pack system was used to obtain data to allow the calculation of the concentration of gas-phase nicotine, 3-ethenylpyridine, 2-ethenylpyridine, pyridine, myosmine, cotinine, and nicotyrine.

The annular diffusion denuder efficiently collected gas-phase nicotine. On the first denuder, nicotine concentrations were $79 \pm 4\%$ of the total gas-phase nicotine collected on both denuders. The total efficiency for the collection of nicotine was 95 ± 5 % for the two denuder sections.

A passive sampling device can reliably collect gas-phase nicotine and 3-ethenylpyridine for samples collected in a chamber.

The chamber experiment results indicated that the following gas-phase compounds may be unique to environmental tobacco smoke in an indoor environment and may be suitable tracers of tobacco smoke:

1. Nicotine
2. 3-Ethenylpyridine
3. Myosmine
4. Nitrous acid
5. Pyridine

Effects of 26 Activities

A controlled study was made by Wallace et al.[4] to determine the effects of each of 26 activities on personal exposure, indoor air concentrations, and exhaled breath for volatile organic chemicals.

The activities engaged in by participants in the study were

1. Tobacco smoking
2. Passive smoking
3. Painting
4. Repairing auto
5. Driving auto
6. Pumping gasoline
7. Mowing lawn
8. Using insecticide
9. Using insect repellant
10. Visiting photograph developing shop
11. Using combustion device
12. Using felt-tip marking pen
13. Using opaquing fluid
14. Showering
15. Tub bathing
16. Swimming
17. Boiling water
18. Washing clothes
19. Washing dishes
20. Using humidifier
21. Cleaning house
22. Polishing floor
23. Polishing furniture
24. Using room air deodorizer
25. Using toilet bowl deodorizer
26. Visiting dry cleaners

For personal air sampling (indoor and outdoor sampling), approximately 9 L of air was drawn through a bed of Tenax GC in a glass tube. At two homes, three co-located samples of 7, 14, and 21 L were collected during each monitoring period. For each participant at each visit to the participant's residence, breath samples were collected. To collect the breath sample, each participant inhaled humidified ultrapure air through a spirometer into a 40-L Tedlar holding bag. An 8.5-L (approximate) aliquot from the holding bag, pumped through a Tenax GC cartridge to concentrate the organic compounds, was used in later analysis. Smokers were requested to wait to smoke their first cigarette each morning until after the breath sample had been collected.

Samples were thermally desorbed from the Tenax GC at 260°C, with a nominal helium flow into a liquid-nitrogen-cooled nickel-capillary trap. By ballistic heating of the nickel trap at 250°C, the condensed vapors were then introduced into a high-resolution fused silica capillary chromatography column.

The constituents of the sample were identified and quantitated by electron mass impact mass spectrometry, by measuring the intensity of the extracted ion current profile.

Four aromatic compounds:

1. Benzene
2. *m* + *p*-Xylene
3. *o*-Xylene
4. Ethylbenzene

are smoking related. Personal monitors did not provide a true indication of exposure to these compounds.

The breath levels of benzene and styrene in cigarette smokers' breath were about five to ten times the level of nonsmokers or pipe and cigar smokers. The major source of exposure to benzene and styrene is mainstream tobacco smoke.[5] In the cigarette smokers' homes, indoor air levels of these two compounds were slightly elevated.

In detecting exposures from smoking that otherwise would not have been detected, measurements of exhaled breath were useful.

ANALYSIS OF GAS PHASE ORGANIC COMPOUNDS IN ULTRA-LOW-TAR DELIVERY CIGARETTE SMOKE

A method "well-suited to the analysis of gas phase organic compounds in ultra-low-tar delivery cigarette smoke" was described by Higgins et al.[6] In the method, the cigarette was smoked directly through a filter and a Tenax GC trap in series.

Thermal desorption programmed-temperature gas chromatography was used to analyze the gas phase organic compounds collected on the Tenax GC resin. To quantitate the collected compounds, external standards were used. The method is particularly applicable to ultra-low-tar cigarettes, however, it can be applied to higher-tar-delivery cigarette products also.

Commercial varieties of cigarettes obtained from domestic and foreign markets, and 1R1 Kentucky Reference cigarettes obtained from the University of Kentucky were used in the development of the method. The cigarettes were stored at −20°C in sealed plastic bags and were conditioned for at least 48 hours at standard conditions of 24 ± 1°C and 60 ± 2% relative humidity before use.

The Tenax GC traps were heavy-wall Pyrex desorption tubes containing 220 ± 15 mg of 35/60 mesh Tenax GC which was pretreated and held in place by solvent-washed and thermally conditioned glass wool. A tapered end of the trap fit a gas chromatograph injection port. The trap was connected between the filter and a single-port smoking machine.

The gas chromatography capillary was 66 m long with the wall coated with 0.21% UCON 50 HB 660, which was prepared by the method described by Higgins,[7] except for a different etching procedure.

The entire trap was used for thermal desorption analysis for 0.2 mg or less of tar. For cigarettes delivering more than 0.2 mg of tar, the Tenax GC was unloaded into a tarred vial in a clean box. Portions of this well-mixed sample, or of a homogeneous dilution made with a known amount of clean Tenax-GC, were to be used for analysis by thermal desorption gas chromatography.

The components of the compounds desorbed from the Tenax GC were detected by flame ionization in the gas chromatograph, were identified by gas chromatography-mass spectrometry (GC-MS), and were verified by comparison of the gas chromatograph retention times with authentic standards where available. External standards were used to quantitate the components.

Since the method could be used to analyze the entire collected sample, it was sufficiently sensitive for analysis of some ultra-low tar cigarette products. Following simple dilution of the Tenax GC, gas samples collected from higher-tar-delivery cigarettes could also be analyzed.

Methane, ethane, propane, ethylene, and many other low molecular weight compounds in tobacco smoke were not retained by the Tenax GC. This was a limitation of the Tenax GC trapping method. That substances breaking through the Tenax GC could be trapped on a carbonaceous material such as Ambersorb XE-340 and subsequently recovered by thermal desorption was suggested by preliminary work.

For 4 ultra-low-tar cigarettes having tar deliveries of from less than 0.01 to 3 mg tar per cigarette, deliveries of 34 gas phase components were determined. For 5 cigarettes delivering 7 to 45 mg tar per cigarette, deliveries of 24 gas phase components were determined.

TOBACCO-SPECIFIC N-NITROSAMINES

Gas chromatography (GC) and a thermal energy analyzer (TEA) have been used in analytical studies to identify the following seven tobacco-specific N-nitrosamines (TSNA) in tobacco and tobacco smoke:[8]

1. N'-nitrosonornicotine
2. N'-nitrosoanatabine
3. N'-nitrosoanabasine
4. 4-(Methylnitrosamino)-1-(3-pyridyl)-1-butanone
5. 4-(Methylnitrosamino)-1-(3-pyridyl)-1-butanal
6. 4-(Methylnitrosamino)-4-(3-pyridyl)-1-butanol
7. 4-(Methylnitrosamino)-4-(3-pyridyl)butyric acid

The TSNA are formed from tobacco alkaloids.

The thermal energy analyzer used was a chemiluminescence detector which was very specific for nitrosamines and was very sensitive.[9] The gas chromatograph had a glass column packed with 10% UCW-982 on Gas Chrom Q. The dimensions of the column was 12 ft × 0.25 in. (2 mm I.D.). The GC was used for separation for snuff tobacco and cigarette smoke.

A GC packing of 3% XE-60 on Gas Chrom Q was used for separation of the major TSNA.

The levels of the seven TSNA identified ranged from 6 to 530 ng per cigarette for tobacco smoke and from 0.01 to 92 ppm for tobacco.

BENZENE IN SMOKING

Introduction

Benzene is produced in the largest volume of any chemical that has been causally linked to cancer in humans.[10,11] In addition, it is a by-product of various combustion processes such as burning of wood and forest fires, of cigarettes, of garbage, and of organic wastes.[12-14]

Potential sources of benzene in the environment include:[15,16]

1. Benzene production
2. Chemical spills
3. Coking of coal
4. Combustion of gasoline
5. Gasoline refining
6. Leaking storage tanks
7. Oil spills
8. Production of benzene-based chemicals
9. Solvent use

Approximately 8.5×10^9 kg of benzene is emitted annually in the U.S. alone.[17]

The Total Exposure Assessment Methodology (TEAM) studies found that benzene levels in air people breathe (personal air) averaged two times higher than concentrations in outdoor air.[18-23] Exposure to benzene concentrations indoors was found to be greater than exposure to the benzene level near gasoline stations, in most cases.[16]

"Smoking is by far the largest anthropogenic source of background human exposure to benzene." [10] It has been reported that smokers had benzene levels in expired air two to ten times higher than those of nonsmokers.[23] Also, nonsmokers who lived with smokers or came in contact with smokers had elevated levels of benzene in their breaths.[23]

Environmental Partitioning of Benzene

The environmental partitioning of benzene was evaluated by Hattemer-Frey et al.[10] They also identified the major sources of human exposure to benzene.

The results of the study showed that 99% of the benzene released into the environment partitions mainly into air. Less than 1% of the benzene released partitions into biota, sediment, soil, suspended sediment, and water.

Inhalation was found to be "the primary route of human exposure to background levels of benzene in the environment." Again, "smoking is by far the largest anthropomorphic source of background human exposure to benzene." About three times more benzene is taken in daily by average smokers (20 cigarettes per day) from smoking than from their exposure to background benzene contamination.

Estimate of Benzene Absorption from Cigarette Smoke

The adsorbed dose of benzene for cigarette smoking was estimated by Travis et al. [24] using a pharmacokinetic model.

The estimation of benzene absorbed from cigarette smoking was based on a pharmacokinetic analysis of the California and New Jersey Total Exposure Assessment Methodology study, TEAM.[22,25] Smokers within the study were grouped as light (less than 10 cigarettes per day), moderate, and heavy (more than 30 cigarettes per day).

For smokers within the TEAM study, the expired air data indicated that the absorbed dose of benzene from cigarette smoke is approximately 40 µg per cigarette. Based on an empirical measurement of 57 µg of benzene per cigarette and the data from the TEAM study, it was predicted that one-half of the total U.S. population exposure to benzene results from active cigarette smoking.[26] This measured value of benzene exposure was of the *release* of benzene during cigarette combustion. The 40 µg per cigarette was an estimate of the amount of benzene *absorbed* per cigarette. The concentration of benzene in the breath of smokers in the TEAM study was consistent with measurements of benzene in mainstream cigarette smoke.[26]

NICOTINE

Intercomparison of Sampling Techniques for Nicotine in Indoor Environments

An intercomparison of sampling techniques for nicotine in indoor environments was made by Caka et al.[27] The sampling systems studied used filter packs, annular denuders, sorbent beds, and passive samplers. The intercomparison of nicotine determinations evaluated the precision and equivalency of each of the techniques. Determinations were made of both airborne gaseous nicotine and particle-phase nicotine.

The four laboratories participating in the study were

1. A group from Brigham Young University (BYU)
2. A group from Harvard University (HAR)

3. A group from the R. J. Reynolds Company (REY)
4. A group from the University of Massachusetts Medical School and Yale University (UM/Y)

The group from BYU used four sampling techniques:

1. The annular diffusion denuder (BYU AD), which determined gas- and particulate-phase nicotine by collection of gas-phase nicotine in an acid-coated diffusion denuder, followed by collection of particulate nicotine. The results gave gas-phase, particulate-phase, and total nicotine concentration.
2. The filter pack (BYU FP) sampling system, consisting of a 47-mm Teflon filter followed by two benzenesulfonic acid (BSA)-coated glass fiber filters. The BSA-coated filters collected gas-phase nicotine as well as any nicotine lost from particles collected on the first filter. Data generated by the filter pack included gas-phase nicotine, particulate-phase (assumed to be the nicotine collected on the Teflon filter), and total nicotine.
3. The passive sampling device (BYU PAS), which collected gas-phase nicotine on a BSA-coated glass fiber filter.
4. The Tenax semi-real-time sampler, consisting of two glass microtubes in sequence. The first tube contained 3 mg of silanized glass wool to collect particulate nicotine. The second tube contained a 20-mg bed of 35/60 mesh Tenax sorbent to collect gas-phase nicotine.

The group from Harvard used two sampling systems:

1. The miniannular denuder (HAR AD), a personal annular denuder including a Teflon-coated glass impactor, which served as a size-selective inlet, and a Teflon filter pack to collect particulate nicotine. The expected efficiency of the denuder system for collection of nicotine was 98.6%.
2. The Millipore cassette (HAR FP), consisting of a Teflon filter followed by a citric acid (CA)-coated glass fiber filter. Gas-phase nicotine and any nicotine lost from particles was collected on the acid-coated filter. Particulate-phase nicotine was collected on the Teflon filter.

The group from the R. J. Reynolds Company used two sampling systems:

1. The XAD-IV sampler (REY XAD), consisting of a 7-cm × 0.6-cm glass tube containing two sections of 20/40 mesh XAD-IV sorbent (80 mg in the primary section and 40 mg in the backup section) separated by a glass wool spacer. The data from the sorbent sections allowed determination of the efficiency of the XAD-IV sorbent bed for the collection of gas-phase nicotine.
2. The passive sampling device (REY PAS), consisting of a stainless steel diffusion sampler.

The group from the University of Massachusetts Medical School and Yale University used two sampling systems:

1. The active sampler (UM/Y FP), a polystyrene Millipore sampling cassette containing a 37-mm Teflon-coated glass fiber filter followed by a Teflon-coated glass fiber filter coated with an aqueous 4% (weight/weight) sodium acid sulfate solution. Gaseous nicotine was collected on the acid-coated filter. Particulate-phase nicotine was collected on the first filter, gas-phase nicotine was collected on the acid-coated filter.

2. A passive sampling device (UM/Y PAS), with a collection surface consisting of a sodium acid sulfate-coated, Teflon-coated glass fiber in a 37-mm-diameter polystyrene air sampling Millipore cassette protected by a PTFE filter windscreen.

Six experiments were conducted in a chamber facility in which environmental tobacco smoke was generated by four smokers at a rate of one cigarette smoked every 10 min. The nicotine rating of the commercial cigarette smoked was 1 mg and the FTC tar rating was 17 mg. The fraction of air exchanged in the chamber facility was changed in order to vary nicotine concentrations to simulate typical indoor environments.

Two different expected concentrations of nicotine were used in the study:

1. A lower expected concentration of nicotine (200 nmol/m^3 or 30 µg/m^3) represented the higher concentrations of nicotine found in indoor environments in which there was smoking. The environments included homes, work environments, and restaurants.

2. A higher expected concentration of nicotine (800 nmol/m^3 or 125 µg/m^3), which had been reported in indoor environments but had seldom been seen.

The amount of nicotine collected was varied by varying sampling time.

Analyses of the collected samples was done in the four groups' respective laboratories.

Determinations of total nicotine using the various sampling systems were generally in good agreement. There was agreement among samplers for determinations of nicotine in the gas phase since more than 95% of the nicotine was present in the gas phase. Surface passivation of stainless steel samplers helped minimize the problem of low results due to adsorption of nicotine by the sampler. Determinations of gas-phase nicotine was not significantly affected by volatilization of particulate-phase nicotine.

Since only 2 to 3% of total nicotine was present as particulate-phase nicotine, the precision of particulate-phase nicotine data was poor. Also, the study indicated that loss of particulate-phase nicotine to the gas phase occurs in a filter pack sampling system. Nicotine was lost from particles during sampling using the studied systems.

Thermal-Desorption-Based Personal Monitoring System

The development of a thermal-desorption-based personal monitoring system for nicotine using Tenax GC as the adsorber was discussed by Thompson et al.[28]

The system was evaluated in controlled environmental tobacco smoke environments in chambers and in offices. The system was also validated in the field.

The air-sampling cartridges were 16-cm sections of 0.25-in. O. D. borosilicate glass tubing. The tubing had been pretreated with ammonium hydroxide by immersing it in 15% ammonium hydroxide overnight and then air drying. The tubing sections were then fire polished on both ends and packed with approximately 200 mg of 35/60 mesh Tenax GC. The packed cartridges thus formed were conditioned at 250°C before use.

In most experiments, personal sampling pumps were used for collection of samples.

Solution standards containing nicotine were prepared by diluting redistilled 98% nicotine in ethyl acetate containing 0.01% triethylamine.

Analyses of collected samples were made on a gas chromatograph equipped with a nitrogen/phosphorus detector (GC-NPD). The parameters for the gas chromatograph were

1. Column — 2 m × 2 mm I.D. glass column
2. Column packing — 10% Carbowax 20 M/2% KOH on 80/100 mesh Chromosorb W-AW
3. Carrier gases — helium, hydrogen, and air
4. Flow rates of carrier gases — 40 mL/min for helium, 4.5 mL/min for hydrogen, 175 mL/min for air
5. Temperature settings — injector, 250°C; detector, 250°C
6. Column oven temperature — initially at 70°C for 8 min, programmed at a rate of 46°C/min to 175°C for 4 min

At these settings, the elution time was 13.4 min for nicotine and 14.0 min for quinoline.

In the development of the system, experiments were performed in chambers and in an unoccupied office. Initial field evaluations were made at the Oak Ridge National Laboratory in work areas, offices, common areas, and dining areas.

Field sampling was done in 36 restaurants in Knoxville, TN and at each of three food courts in shopping malls. In the restaurants, the following information was recorded:

1. Number of smokers
2. Number of cigarettes, cigars, and pipes observed to have been smoked
3. Distance to the closest observed smoker
4. Restaurant volume
5. Crowd density
6. Type of meal served (lunch or dinner)

During the time of sampling, the information was recorded on a sampling data sheet. No attempt was made to determine the air exchange rate within the sampled facility. Also, no determination was made of the number of smokers smoking at any time and no determination was made of the smoker turnover.

It was considered that the major factors likely to affect nicotine concentrations in a public location included:

1. Number of cigarettes smoked
2. Time required for smoking
3. Proximity of the smokers to the sampling location
4. Volume of the room
5. Air exchange rate

The thermal-desorption-based personal monitoring system for nicotine, using Tenax GC as the adsorption material, had a low detection limit and was unobtrusive in use.

Results for field determination of nicotine concentration in the restaurants compared favorably with those obtained by other researchers using different methods for sampling in restaurants and other public places.

When trace quantities of nicotine are being analyzed or processed, the development studies indicated the need for the use of a basic compound for desorption enhancement or sample modification.

Passive Diffusion Monitor for Nicotine

In a preliminary note, Hammond and Leaderer[29] reported initial results in the designing of a passive monitor to measure personal exposure to ETS on the basis of diffusion of nicotine.

A modified 37-mm-diameter polystyrene air sampling cassette was used in the construction of the passive monitor. A support and a treated filter were inserted in the bottom of the cassette. The treated filter was a glass fiber coated with Teflon. It was treated by saturation with an aqueous solution of 4% sodium bisulfate and 5% ethanol. After saturation, the filter was allowed to dry. The filter was held in place by a windscreen.

Under controlled conditions, the monitors were tested in an environmental chamber. Under conditions of constant smoking and ventilation and at 5 concentrations of ETS, 5 sets of 10 monitors each were tested for 4 to 5 hours. ETS concentrations ranged from 150 to 1500 $\mu g/m^3$ total particulate mass. The empirical sampling rate, 24 mL/min, agreed well with the theoretically calculated rate of 25 mL/min.

The empirical sampling rate was determined from the slopes of plots of micrograms of nicotine collected by the passive monitors (during each test) against the product of the nicotine concentration and the number of hours the passive monitors were exposed to cigarette smoke. Nicotine concentration was determined using active samplers.

When the sampler was exposed to conditions found in the "real" world, intermittent and varying concentrations of ETS over several days, the passive monitors collected the same amount of nicotine as the smoking samplers that were exposed only for the actual 4- to 5-hour smoking periods.

The passive monitor could measure exposure to low levels of airborne nicotine, as a proxy for environmental tobacco smoke. The monitor could measure a wide range of ETS concentrations.

Sampling of Indoor Air for Nicotine Using Graphitized Carbon Black in Quartz Tubes

A new adsorbent material for sampling and a different thermal desorption principle for the determination of nicotine in ambient air was reported by Vu-Duc and Huynh.[30] Generation of nicotine in the vapor phase and sampling of sidestream tobacco smoke in an experimental chamber were used to validate the method.

The samplers consisted of transparent quartz tubes filled with 70 mg of 20/40 mesh Carbotrap graphitized carbon black. The tubes were plugged at both ends with silanized quartz. Before use, the tubes were desorbed by three thermal desorption intervals of 10 sec each.

A microwave thermal desorber and a gas chromatograph were used for analyses. The desorption conditions were

1. Interface temperature — 200°C
2. Desorption time — 4 sec
3. Desorption power — "Low"

A Cr-Alumel thermocouple probe inserted in the sorbent bed was used to measure the temperature produced by the thermal desorber. The gaseous effluent from the tubes was introduced directly into the injector of the gas chromatograph.

The parameters for the gas chromatograph were

1. Mode — split, 1.5:10
2. Column — 25 m × 0.25 I.D. WCOT fused-silica capillary column
3. Column coating — 0.25-µg film of CP-Sil-5CB (Chrompack)
4. Detector — thermionic nitrogen-specific detector (NPD)
5. Injector temperature — 250°C
6. Oven temperature — 120°C isothermal
7. Carrier gas — nitrogen
8. Manometer pressure of carrier gas — 14 psi
9. Detector temperature — 280°C
10. Gas supply to detector — 4 mL/min hydrogen, 30 mL/min nitrogen (as the makeup gas), and 175 mL/min air

Nicotine was generated in the vapor phase for the low concentration range, less than 1 µg/L, by the permeation process[31] to test the performances of the Carbotrap tubes and of the overall analytical procedure.

The procedure was applied to the determination of nicotine at various concentrations in an experimental chamber in sidestream tobacco smoke. Under the experimental conditions, the limit of detection of the method was found to be 25 ng nicotine per tube. The method was applicable to a 4-hour sampling time at 0.1 L/min or to a 30-min period at 1 L/min at the low concentration end of 1 $\mu g/m^3$ nicotine.

ASTM Standard Test Method for Nicotine in Air

The American Society for Testing and Materials, ASTM, (Committee D-22 on Sampling and Analysis of Atmospheres, Subcommittee D22.05 on Indoor Air) has produced a standard test method for the determination of nicotine in air.[32] Nicotine is collected by adsorption on a sorbent resin, extracted from the sorbent, and determined by gas chromatography with nitrogen-selective detection.

The method is summarized as follows:

"A known volume of air is drawn through a sorbent sampling tube containing XAD-4 resin to absorb the nicotine present. The XAD-4 resin from the the two tube sections are (sic) each transferred to a 2 mL autosampler vial and the nicotine is desorbed with ethyl acetate containing 0.01% triethylamine. A known quantity of quinoline is added as an internal standard. An aliquot of the desorbed sample is injected into a gas chromatograph equipped with a thermionic specific (nitrogen-phosphorus) detector. The area of the resulting nicotine peak is divided by the area of the internal standard peak and compared with area ratios obtained from the injection of standards."

Capillary Chromatographic Analysis for Nicotine

Capillary chromatographic analysis has been used for nicotine.[33] Sampling for nicotine was done in office buildings and in a small chamber.

Cambridge filters, saturated with an aqueous solution of 4% sodium bisulfate, were allowed to dry and then used for nicotine analysis. The procedure of Hammond et al.[34] was used to prepare the filters. The sampling chamber was loaded with smoke using a smoking machine,[35] the filters were used to collect nicotine from air sampled at a flow rate of 1 L/min using personal sampling pumps. Nicotine was desorbed from the filters using the method of Hammond et al.[34]

The nicotine analysis was made using a gas chromatograph (GC) with a nitrogen-phosphorus detector (NPD). The operating parameters for the GC were

1. Column — DB-5 30 m × 0.25 mm I.D. column
2. Column coating — 0.25-μg film
3. Injection — splitless with the septum purge off during injection until 0.5 min
4. Oven temperature — 120°C
5. Injection temperature — 260°C
6. Detector temperature — 260°C
7. Detector gas flow rates — 3.0 mL/min for hydrogen, 30 mL/min for helium, 80 mL/min for air

8. Carrier gas — helium
9. Carrier gas average linear velocity — 50 cm/sec (equivalent to 1.6 cm³/min)

To indicate the presence of environmental tobacco smoke, nicotine analysis using the capillary GC coupled with the NPD was used. Separation of the nicotine from other nitrogen-containing compounds in ETS required high-resolution GC. The NPD had been found in previous work[36] to be particularly useful for nicotine analysis in studies of exposure to passive tobacco smoke.

Nicotine as a Tobacco Smoke Indicator in Mechanically Ventilated Office Towers

Goyer[37] reported on studies of air quality, with respect to chemical contaminants, in 17 mechanically ventilated office towers located in urban areas of Quebec. The building heights ranged from 3 to 30 stories and were less than 25 years old. Contaminants most likely to be present in indoor air in detectable concentrations were measured at emission sources and at work stations. The contaminants measured included:

1. Airborne dust
2. Volatile organic compounds
3. Nicotine
4. Carbon monoxide
5. Carbon dioxide
6. Formaldehyde
7. Ozone
8. Radon

Measurements were made over periods of up to 2 weeks, in the winter season. Indoor air was sampled at various times of the day—during working hours and outside of working hours.

Nicotine was chosen as the indicator of tobacco smoke because its concentration in air is fundamentally dependent on smoking behavior. The concentration of nicotine decays rapidly after emission. Nicotine can be adsorbed on particulates.

Nicotine and dust were found to be closely related to occupant activity, and they were well correlated.

PROXIES, SURROGATES, TRACERS, AND MARKERS

Genotoxic Components and Tracers

The concentration of a number of genotoxic components as well as potential tracers of environmental tobacco smoke under controlled and environmental conditions was studied by Lofroth et al.[38]

Carbon monoxide, nitrous oxides, aldehydes, alkenes, and particulate matter were measured. Nicotine was measured as a tracer of environmental tobacco smoke, with the assertion that nicotine was the best available tracer. Indoor experiments on the measurement of nicotine were performed in a chamber at the U.S. Environmental Protection Agency facility at the University of North Carolina in Chapel Hill. Measurements were also made in a tavern.

The laboratory tests were performed in a ventilated 13.6-m³ Plexiglas chamber. Research cigarettes, which had been equilibrated at 22°C and 60% relative humidity for 48 hours, were smoked by machine. Nicotine was collected on filters impregnated with bisulfate. In the first series of tests, the filters were placed downstream from both personal samplers and Anderson samplers. The nicotine was extracted and analyzed by gas chromatography by procedures described by Hammond et al.[34]

In a local tavern, two studies were made. Nicotine was collected on a Teflon-coated glass fiber filter, and on a filter impregnated with bisulfate. The sampling rate, using an Anderson sampler, was 20 L/min.

In the chamber tests, the concentrations and yields of nicotine were lower in the first series than in the second series. The nicotine yield in the first series was 800 µg/cigarette. The nicotine yield in the second series was 3300 µg/cigarette. The chamber contained more adsorbant surfaces in the first series than in the second series. In the first series, the relative humidity was approximately 50 to 60%. In the second series, the relative humidity was approximately 30%. Although surface area is likely to alter surface deposition, Leaderer[39] had shown that relative humidity also affects surface deposition.

Following smoking of one or two cigarettes by a smoker in the chamber, ETS from two brands of commercial cigarettes was analyzed. The composition of the smoke was found to be similar to that from research cigarettes.

In air sampled during normal smoking in the tavern, all ETS components analyzed were found to be highly elevated in the indoor environment. The concentrations of analyzed components were not found to be conspicuously much higher or lower relative to each other than studies of airborne yield of research cigarettes would have been expected to exhibit.

Isoprene, among the unsaturated hydrocarbons, showed promise as a tracer of tobacco smoke. Studies of other potential indoor sources of isoprene that could interfere with it as a tracer of tobacco smoke are needed.

Identification of a Group of Potential Tracers of ETS

The objective of a study by Benner et al.[40] was the detailed chemical characterization of both the gas-phase and the particulate-phase constituents of environmental tobacco smoke in order to identify a group of potential tracers that would meet the National Academy of Sciences recommended criteria:

1. Uniqueness
2. Ease of measurement

3. Similarity in emission rates for different tobaccos
4. Consistent ratios to ETS compounds of interest

The characterization of particulate-phase ETS was described and recommendations were made of several potential tracers which were identified.

Acid-washed, fired quartz filters were used to collect samples for detailed organic analysis. The collection was made either 15 to 30 min or 4 hours after cigarette combustion had occurred. The filter was then Soxhlet extracted with 100 mL of a dichloromethane-methanol (1:1) mixture for 36 hours. The resulting solutions were concentrated to near dryness by rotary evaporation and were saponified with 20 mL of aqueous 6 N KOH for 48 hours at room temperature. Dichloromethane (3×25 mL) liquid-liquid extraction was used to recover neutrals and bases from the basic aqueous solution. After acidification with HCl, the acid fraction was recovered by extraction with CH_2Cl_2($3 \times$ 25 mL).

Identification of potential conservative organic tracer species for the particulate phase of ETS was made by analyzing the resulting fractions by gas chromatography-mass spectrometry (GC-MS). Quantitative determinations were made using a GC-flame ionization detector (FID) and GC-nitrogen/phosphorus detection (NPD) on a gas chromatograph.

Based on the experimental results, the following particulate-phase components of ETS were proposed as possible tracers: (1) nicotine and related compounds, (2) solanesol, and (3) sterols and sterenes.

Other particulate-phase constituents of ETS suggested for monitoring to determine their relationship to major ETS compounds include potassium and respirable suspended particles. The average concentration of potassium in ETS particles was 240 µmol/g (1 wt%). The ratio of respirable suspended particles (RSP) to carbon monoxide (CO) was found to be quite constant with the number of cigarettes smoked, averaging 4.2 ± 0.8 g of RSP per mole of CO.

Solanesol as a Tracer of ETS in Indoor Air

Results of experiments which demonstrated that solanesol, a primary terpenoid alcohol, appeared to be well suited as a sensitive, unambiguous tracer of environmental tobacco smoke particles in indoor environments, were presented by Ogden and Maiolo.[41]

Fused silica capillary tubing was used in column preparation. Solanesol, N-Obis(trimethylsilyl)trifluoroacetamide, with and without 1% added trimethylchlorosilane, and 1-triacontanol were used for sample/standard preparation. Fluoropore filter pads were used for sample collection in conjunction with personal sampling pumps. Sorbent tubes containing XAD-4 resin were used to collect nicotine and to verify the absence of solanesol in the vapor phase of ETS aerosol.

Analyses were performed on gas chromatographs equipped with on-column injectors and autosamplers with either flame ionization detection (FID) or mass spectrometric detection in the selected ion mode.

After sampling and gravimetric determination of respirable suspended particles, the filter pads were transferred from their cassette holders to 1-mL reaction vials, spiked with 20 µL of internal standard solution, and then covered with 0.75 mL of pentane. The vials were capped and placed in an ultrasonic bath at 37°C for 30 min. The filters were then removed, the pentane was evaporated, and derivatization was performed.

Solanesol, to be a reliable tracer of ETS particulate matter (ETS-PM), must be in a constant proportion to the particulate fraction of ETS aerosol for a variety of smoking products and environmental conditions. To study the relationship of solanesol to ETS-PM, environmental tobacco smoke was generated in an 18-m^3 environmental test chamber operated at 23°C and 50% relative humidity in the static mode (no air exchanges) with recirculating fans on. Cigarettes were "human smoked" in the chamber at 1 puff/min and samples were collected at 2 L/min for 2 hours.

For the experimental data, the limit of detection was estimated to be 0.2 µg/m^3 for 2-hour sample duration at 2 L/min. Solanesol was indicated to be 2 to 3% by weight of respirable suspended particles (RSP) attributable to ETS for commercial cigarettes. Consequently, the solanesol/RSP weight ratio could be used to apportion total RSP into ETS and non-ETS contributions. This approach was used to correctly predict the ETS contribution to a mixture of RSP from cigarette, candle, and oil lamp sources with an error of 10%.

Ogden and Maiolo[41] concluded that of all the potential tracers that had been suggested for quantifying ETS particulate concentrations in indoor environments, solanesol appeared to be the best candidate; although more work needed to be done before solanesol could be used as a routine tracer of ETS.

It was anticipated that the only measurable contribution of solanesol, like nicotine, to an indoor environment would be from tobacco sources. However, unlike nicotine, solanesol was not expected to shift equilibrium between vapor and particle phases of the ETS aerosol under any normal conditions encountered in an indoor environment.

Quinoline and Isoquinoline as Markers for ETS

Chuang et al.[42] did a pilot field study to measure the concentrations of polycyclic aromatic hydrocarbons (PAHs), PAHs derivatives, and nicotine in air in selected residences. They found that ETS was the most significant influence on indoor pollution levels, and that there were good correlations between nicotine and quinoline and between nicotine and isoquinoline. They recommended that quinoline and isoquinoline, instead of nicotine, be used as ETS markers.

REFERENCES

1. "The Health Consequences of Involuntary Smoking," A Report of the Surgeon General. U.S. Department of Health and Human Services, 1986.
2. *Environmental Tobacco Smoke. Measuring Exposures and Assessing Health Effects,* National Research Council. Washington, DC: National Academy Press, 1986.
3. Eatough, D. J., C. L. Benner, J. M. Bayona, G. Richards, J. D. Lamb, E. A. Lewis, and L. D. Hansen. "Chemical Composition of Environmental Tobacco Smoke. 1. Gas-Phase Acids and Bases," *Environ. Sci. Technol.* 23:679–687 (1989).
4. Wallace, L. A., E. D. Pellizzari, T. D. Hartwell, V. Davis, L. C. Michael, and R. W. Whitmore. "The Influence of Personal Activities on Exposure to Volatile Organic Compounds," *Environ. Res.,* 50:37 (1989).
5. Wallace, L. A., E. D. Pellizzari, T. Hartwell, K. Perritt, and R. Zigenfus. "Exposure to Benzene and Other Volatile Organic Compounds from Active and Passive Smoking," *Arch. Environ. Health* 42:272–279 (1987).
6. Higgins, C. E., W. H. Griest, and G. Olerich. "Application of Tenax Trapping to Analysis of Gas Phase Organic Compounds in Ultra-Low Tar Cigarette Smoke," *J. Assoc. Off. Anal. Chem.* 66:1074–1083 (1983).
7. Higgins, C. E. "Preparation of Polar Glass Capillary Columns," *Anal. Chem.* 53:732 (1981).
8. Brunnemann, K. D., and D. Hoffmann. "Analytical Studies on Tobacco-Specific N-Nitrosamines in Tobacco and Tobacco Smoke," *Crit. Rev. Toxicol.* 21:235–240 (1991).
9. Fine, D. H., F. Rufeh, D. Lieb, and D. P. Rounbehler. "Description of the Thermal Energy Analyzer (TEA) for Trace Determination of Volatile and Non-Volatile N-Nitroso Compounds," *Anal. Chem.* 47:1188 (1975).
10. Hattemer-Frey, H. A., C. C. Travis, and M. L. Land. "Benzene: Environmental Partitioning and Human Exposure," *Environ. Res.* 53:221–232 (1990).
11. U. S. Environmental Protection Agency (U.S. EPA), "National Emissions Standards for Hazardous Air Pollutants: Regulation of Benzene," *Fed. Reg.* 49(110):23,478–23,495 (1984).
12. Fishbein, L.. "An Overview of Environmental and Toxicological Aspects of Aromatic Hydrocarbons. I. Benzene," *Sci. Total Environ.* 40:189–218 (1984).
13. International Agency for Research on Cancer (IARC), "IARC Monograph on the Evaluation of the Carcinogenic Risk of Chemicals to Humans: Some Industrial Chemicals and Dyestuffs," 29:93–148 (1982).
14. Webster, R. C., H. I. Mailbach, L. D. Gruenke, and J. C. Craig. "Benzene Levels in Ambient Air and Breath of Smokers and Nonsmokers in Urban and Pristine Environments," *J. Toxicol. Environ. Health* 18:567–573 (1986).
15. Agency for Toxic Substances and Disease Registry (ATSDR),"Toxicological Profile for Benzene" (Draft), U. S. Public Health Service, Atlanta, GA (1987).
16. U. S. Environmental Protection Agency (U.S. EPA), "Benzene: Occurrence in Drinking Water, Food, and Air," Science and Technology Branch, Criteria and Standards Division, Office of Drinking Water, Washington, DC (1983).
17. SRI International, "Directory of Chemical Producers U. S., 1987," (1988).

18. Hartwell, T. D., E. D. Pellizarri, R. L. Perritt, R. W. Whitmore, H. S. Zelon, L. S. Sheldon, C. M. Sparacino, and L. Wallace. "Results from the Total Exposure Assessment Methodology (TEAM) Study in Selected Communities in Northern and Southern California," *Atmos. Environ.* 21:1995–2004 (1987).

19. Hartwell, T. D., E. D. Pellizarri, R. L. Perritt, R. W. Whitmore, H. S. Zelon, and L. Wallace. "Comparison of Volatile Organic Levels Between Sites and Seasons for the Total Exposure Assessment Methodology (TEAM) Study, *Atmos. Environ.* 21:2413–2424 (1987).

20. Wallace, L. "Personal Exposures, Indoor and Outdoor Air Concentrations of Selected Volatile Organic Compounds Measured for 600 Residents of New Jersey, North Dakota, North Carolina and California," *Toxicol. Environ. Chem.* 12:215–236 (1986).

21. Wallace, L., R. Zweidinger, M. Erickson, S. Cooper, D. Whitaker, and E. Pellizarri. "Monitoring Individual Exposure: Measurements of Volatile Organic Compounds in Breathing-Zone Air, Drinking Water, and Exhaled Breath," *Environ. Int.* 8:269–282 (1982).

22. Wallace, L., E. Pellizarri, T. Hartwell, C. Sparacino, L. Sheldon, and H. Zelon. "Personal Exposures, Indoor-Outdoor Relationships and Breath Levels of Toxic Air Pollutants Measured for 355 Persons in New Jersey," *Atmos. Environ.* 19:1651–1661 (1985).

23. Wallace, L., E. Pellizzari, T. Hartwell, R. Perritt, and R. Ziegenfus. "Exposure to Benzene and Other Volatile Compounds from Active and Passive Smoking," *Arch. Environ. Health* 42:272–279 (1987).

24. Travis, C. C., P. H. Craig, and J. C. Bowers. "Characterization of Human Exposure to Ambient Levels of Benzene Using Pulmonary 'Wash-Out' Data," *Atmos. Environ.* 25A:1643–1647 (1991).

25. Wallace, L., et al. "The California TEAM Study: Breath Concentrations and Personal Exposures to 26 Volatile Compounds in Air and Drinking Water of 188 Residents of Los Angeles, Antioch, and Pittsburg, CA," *Atmos. Environ.* 22:2141 (1988).

26. Higgins, C. E., W. H. Griest, and G. Olerich. "Application of Tenax Trapping to Analysis of Gas Phase Organic Compounds in Ultra-Low Tar Cigarette Smoke," *J. Assoc. Off. Analyt. Chem.* 66:1074 (1983).

27. Caka, F. M., D. L. Eatough, E. A. Lewis, H. Tang, S. K. Hammond, B. P. Leaderer, P. Koutrakis, J. D. Spengler, A. Fasano, J. McCarthy, M. W. Ogden, and J. Lewtas. "An Intercomparison of Sampling Techniques for Nicotine in Indoor Environments," *Environ. Sci. Technol.* 23:429–435 (1989).

28. Thompson, C. V., R. A. Jenkins, and C. E. Higgins. "A Thermal Desorption Method for the Determination of Nicotine in Indoor Environments," *Environ. Sci. Technol.* 24:1196–1203 (1990).

29. Hammond, S. K., and B. P. Leaderer. "A Diffusion Monitor to Measure Exposure to Passive Smoking," *Environ. Sci. Technol.* 21:494–497 (1987).

30. Vu-Duc, T., and Cong-Khanh Huynh. "Graphitized Carbon Black in Quartz Tubes for the Sampling of Indoor Air Nicotine and and Analysis by Microwave Thermal Desorption-Capillary Gas Chromatography," *J. Chromatogr. Sci.* 29:179–183 (1991).

31. Dharmarajan, V., and R. J. Rando. "Dynamic Calibration of a Continuous Organo-Isocyanate Monitor for Hexamethylene Diisocyanate," *Am. Ind. Hyg. Assoc. J.* 41:437–441 (1980).

32. American Society for Testing Materials (ASTM), "Standard Test Method for Nicotine in Indoor Air," D5075–90a, Philadelphia, PA (1990).

33. Bayer, C. W., and M. S. Black. "Capillary Chromatographic Analysis of Volatile Organic Compounds in the Indoor Environment," *J. Chromatogr. Sci.* 25:60–64 (1987).

34. Hammond, S. K., B. P. Leaderer, and A. C. Roche. "Collection and Analysis of Nicotine as a Marker for Environmental Tobacco Smoke," *Atmos. Environ.* 21:457–461(1987).

35. Black, M. S., and C. W. Bayer. "Formaldehyde and Volatile Organic Compound Exposures from Some Consumer Products," *Proceedings IAQ '86, Managing Air for Health and Energy Conservation,* Atlanta, GA (1986).

36. Williams, D. C., J. R. Whitaker, and W. Jennings. "Air Monitoring for Nicotine Contamination," *J. Chromatogr. Sci.* 17:259–261 (1984).

37. Goyer, N. "Chemical Contaminants in Office Buildings," *Am. Ind. Hyg. Assoc. J.* 51:615–619 (1990).

38. Lofroth, G., R. M. Burton, L. Forehand, S. K. Hammond, R. L. Seila, R. B. Zweidinger, and J. Lewtas. "Characterization of Environmental Tobacco Smoke," *Environ. Sci. Technol.* 23:610–614 (1989).

39. Leaderer, B. P., Private communication.

40. Benner, C. L., J. M. Bayona, F. M. Caka, H. Tang, L. Lewis, J. Crawford, J. D. Lamb, M. L. Lee, E. A. Lewis, L. D. Hansen, and D. J. Eatough "Chemical Composition of Environmental Tobacco Smoke. 2. Particulate-Phase Compounds," *Environ. Sci. Technol.* 23:688–699 (1989).

41. Ogden, M. W., and K. C. Maiolo. "Collection and Determination of Solanesol As a Tracer of Environmental Tobacco Smoke in Air," *Environ. Sci. Technol.* 23:1148–1154 (1989).

42. Chuang, J. C., G. A. Mack, M. R. Kuhlman, and N. K. Wilson. "Polycyclic Aromatic Hydrocarbons and Their Derivatives in Indoor and Outdoor Air in an Eight-Home Study," *Atmos. Environ.* 25B:369–380 (1991).

CHAPTER 7

Gasoline Vapor

INTRODUCTION

Gasoline is a mixture of hydrocarbons with a wide boiling point range of compounds.[1] To match the performance characteristics of gasoline with requirements of the environmental conditions and geographic areas in which they are used, gasoline blends are formulated. A monitoring method for measuring exposures to gasoline vapor must be capable of acceptable performance for the entire range of gasoline blends.

Dose-related increases in the incidence of kidney damage and kidney cancer in male rats and in the incidence of liver tumors in female rats exposed to totally vaporized unleaded gasoline have been shown in toxicological studies.[2,3] However, it was hypothesized that occupational exposure of humans to gasoline vapors should be substantially different than that received by the rats.[4,5] A survey of epidemiological studies indicated that data exist suggesting an elevated risk of cancer of the kidney and prostate for certain small groups of workers within the petroleum industry.[6] Although findings for laboratory exposures to whole gasoline are not indicative or representative of typical human exposures,[7] the findings showing some potential health risks to humans have led to the intensification of activities related to industrial hygiene effects of gasoline exposure in North America and Europe.[8]

The American Petroleum Institute (API) and member companies undertook studies to develop better analytical methods for the quantification of gasoline exposure and to assess worker exposures to gasoline vapors.[8] There are no federal standards regulating gasoline vapor in air; however, there is an American Conference of Governmental Industrial Hygienists 8-hour time-weighted average (TWA) Threshold Limit Value (TLV) for gasoline of 300 ppm.[8]

The highest worker exposures to gasoline vapors usually occur during handling of the product (for example, loading operations).[9] The major compounds of the vapor exposure are low boiling point compounds that readily vaporize, even under frigid conditions. These lighter components will be present in a vapor headspace of a storage tank or vehicle fuel tank in relatively high concentrations.[9] Since the specific components of potential health risk have not been identified, health professional must be able to measure exposures to all of the constituents, including low boiling point compounds with carbon numbers below 6. *Carbon number* is defined as the number of carbon atoms in a compound, indicated by a subscript on the letter C.

In this chapter, measurements of gasoline vapor exposure in a variety of workplace situations will be reviewed.

EVALUATION OF A GASOLINE VAPOR SAMPLING METHOD

Russo et al.[1] evaluated a charcoal tube gasoline vapor sampling method used to monitor worker exposures to gasoline vapors. The effects of temperature and humidity on breakthrough volume, as an indicator of performance limits of the method, were particularly studied.

To evaluate the performance of the charcoal sampling tube method for the contaminants to which workers are exposed, the gasoline vapor used was from gasoline equilibrium headspace rather than vapor from volatilization of bulk liquid gasoline. The concentrations of total hydrocarbon were 20 and 100 ppm, at 20°C. Collection efficiency and component breakthrough volume of isobutane were evaluated in two steps:

1. The collection efficiency of three lower boiling target gasoline components (isobutane, *n*-butane, and isopentane) were evaluated at 20°C and less than 40% relative humidity (RH). The sampling flow rates were 10 and 100 mL/min. Storage stability was evaluated after 0, 8, and 14 days at room temperature.
2. Component breakthrough volume of isobutane was evaluated as a function of RH at 32°C.

Aliquots of the headspace vapor above liquid gasoline, at equilibrium at 20°C in a closed container, were used to generate test environments. Each aliquot was removed from the container and introduced into a 950-L stainless steel static test chamber; the temperature of the test chamber was 20°C and the RH was less than 40%. Six sampling ports on the test chamber were used to collect simultaneous duplicate sets of three samples each at flow rates of 10 and 100 mL/min.

Concentration levels in the test chamber of the three target compounds (isobutane, butane, and isopentane) and total hydrocarbons (THC) were verified by injecting 1 mL of the chamber atmosphere into a gas chromatograph (GC). The GC analysis throughout the 7-hour sampling period provided an independent measurement of the hydrocarbon level. The GC was calibrated using vapor standards containing n-hexane and the three target compounds.

The chamber atmosphere samples were drawn through a sorbent train consisting of two charcoal tubes in series: a large (400-mg front section/200-mg back section) charcoal tube, followed by a small (100-mg front section/50-mg back section) charcoal tube. The tubes were butted together and connected with 3/16-in. I.D. Tygon tubing.

Sample breakthrough volume was determined using large (400-mg front section/200-mg back section) charcoal tubes with the 200-mg back sections removed. A flow of 100 mL/min was maintained through each of the large charcoal tubes. The test chamber was maintained at 32.2°C and was humidified to attain desired temperature/humidity conditions. The effluent from the chamber passed through one of two outlets of a Teflon plenum connected to a pump. The other outlet was connected to the sample loop of a GC equipped with a flame ionization detector. The effluent from the pump was diverted into the GC every 15 min for the determination of hydrocarbon breakthrough. The GC was calibrated with isobutane, at 5% of the isobutane concentration in the test chamber.

The gas chromatograph/flame ionization detector used for analysis of the charcoal tubes had the following parameters:

1. Column — 30 ft × 1/8 in. stainless steel packed column
2. Initial oven temperature — 35°C
3. Temperature program — 5°C/min
4. Detector — flame ionization detector
5. Peak resolution — per NIOSH P&CAM 127 Method[10]

Desorption was with 1 mL of carbon disulfide for 30 min for the small tubes, and 2 mL of carbon disulfide for 30 min for the large tubes.

The collection efficiency (% recovery) of the method at 100-ppm vapor concentration at 20°C and relative humidity of less than 40%, and flow rates of 10 and 100 mL/min, was found to be acceptable (greater than 90%) for total hydrocarbons, isobutane, n-butane, and isopentane. At an approximate 20-ppm concentration of the headspace gasoline vapor, under the conditions above, the collection efficiency for total hydrocarbons, isobutane, n-butane, and isopentane was lower, but acceptable based on the NIOSH collection efficiency criteria (greater than 75%).[11] In both cases, all of the vapor was collected in the front section of the large tube.

Samples collected from the 100-ppm test chamber were tested for storage stability. After storage at room temperature for 0, 8, and 14 days, the samples were analyzed. Samples collected at a flow rate of 100 mL/min for 420 min, a total volume of 42.0 L, showed no sample loss but there was migration of isobutane, n-butane, and isopentane. More migration was detected on the tubes stored for 14 days than on those stored for 8 days. There was no migration detected for samples collected at a flow rate of 10 mL/min for 420 min—a total volume of 4.2 L. It was concluded that gasoline vapor samples collected under these conditions could be stored at room temperature for at least 14 days without loss of hydrocarbon, and it was observed that "Although migration does not affect accuracy, it is undesirable because it mimics breakthrough and samples may be assumed erroneously to be invalid."[12]

It was observed that, although water vapor adversely affects collection efficiency, it is possible to obtain accurate monitoring results by optimizing sample size and sample rate to maximize collection efficiency. Also, sample volume limits are flow-rate dependent. Since, for light ends, sample migration occurs during storage and increases with time, analyses of samples should be made as promptly as possible.

Charts relating sampling parameters to environmental conditions are presented to aid the field hygienist in the application of the results of the study.

CHARACTERIZATION OF GASOLINE VAPOR WORKPLACE EXPOSURES

Halder et al.[4] described a characterization of workplace exposures to gasoline vapors in the petroleum industry. Gasoline vapor exposures were monitored for (1) truck drivers and terminal operators from five terminal loading facilities, (2) dockmen and seamen at two tanker/barge loading facilities, and (3) attendants at a single expressway service plaza.

Of the five terminal loading facilities, three were equipped with bottom loading and vapor recovery; one was equipped with bottom loading but no vapor recovery; and the fifth was equipped with top loading but no vapor recovery. Personal exposures at these locations of truck drivers and terminal operators were monitored with a 3M 3500 Organic Vapor Monitor attached to clothing in the breathing zone. Sampling of average duration of 9.2 hours measured hydrocarbon exposures throughout the workday. Over a 12-month period, 183 samples were collected at the terminals. Of these:

1. There were 37 samples that represented the full range of exposures selected for more detailed analysis of composition; of these, 25 samples were analyzed for n-butane, isobutane, n-pentane, isopentane, and various heavier hydrocarbons; 12 samples were extensively analyzed for all components
2. There were 146 samples analyzed for hydrocarbons with more than six carbon atoms, i.e., a carbon number greater than 6.

The sampling at the two tanker/barge loading facilities was part of an industry project sponsored by the American Petroleum Institute.[13] At these two facilities,

11 personal exposure samples (9 from a tanker loading facility, and 2 from a barge loading facility) were collected using 600-mg SKC charcoal tubes connected to pumps. The sampling flow rate was 50 mL/min; the sampling period was of an average 3.8-hour duration, covering the typical workshift period. A contractor performed the analytical work on the samples.

At the expressway service plaza, exposure monitoring was performed for paired gasoline service stations immediately off the exit ramps on either side of a major expressway. Service attendants were monitored in a manner similar to that used for the terminal truck drivers and operators, for an average collection period of 7.2 hour lasting throughout the workday — 21 samples were collected and were analyzed for hydrocarbons with 6 or more carbon atoms only.

In addition to the three monitoring surveys above, a small ancillary monitoring study was conducted at the Amoco Research Center in Naperville, IL to simulate exposures of consumers to gasoline vapors while they are refueling automobiles. During the fueling of test cars, sampling durations varied from 5 to 446 min. During individual sampling periods, the number of cars fueled varied from 1 to 37. Simultaneously, two sampling methods were used: (1) 3M 3500 Organic Vapor Monitor diffusion badges, and (2) 600-mg SKC charcoal tubes connected to a pump; the flow rates varied from 56 to 78 mL/min. Six sample pairs were taken; four sample pairs were hydrocarbons with more than six carbon atoms, the two remaining sample pairs were analyzed for hydrocarbons with more than four carbon atoms.

Throughout the program, exposure results for all samples were adjusted to an 8-hour time-weighted average (TWA, 8-hr). It was assumed that there was no other measurable exposure during periods that were not monitored.

Sampling rates developed by the 3M company were used to calculate n-pentane exposures and exposures to hydrocarbons with more than six carbon atoms where the 3M 3500 Organic Vapor Monitor diffusion badge was used. The sampling rates developed were claimed to be accurate to $\pm 5\%$. For the remaining hydrocarbons with five or six carbon atoms, sampling rates were determined using the 3M Company Sampling Validation Protocol; reference sources provided specific diffusion coefficients for these compounds, with an accuracy estimated to be within $\pm 10\%$. The contractor who analyzed samples collected on SKC charcoal tubes at the marine loading operation used "internal standard calibration prior to sample analysis to determine desorption efficiencies for each component expected to be found in gasoline vapor."

The 37 terminal samples and 2 of 6 automobile refueling exposure samples were analyzed using a gas chromatograph (GC). The parameters for the GC were

1. Column — 60 m long × 0.25 mm diameter capillary column
2. Column packing — fused silica DB-1 exchange bed
3. Carrier gas — hydrogen
4. Carrier gas flow velocity — 40 cm/sec
5. Temperature program — began at –10°C, held for 4 min, then increased at 2°C/min up to 250°C
6. Detector — flame ionization detector

Carbon disulfide was used to desorb the hydrocarbons from the monitoring badges.

The lower limit of detection for all compounds analyzed by this GC-desorption method was 1 μg per sample, with the exception of benzene for which it was 2 μg per sample. Detection of hydrocarbons with more than four carbon atoms was accomplished using this method.

For 146 of the terminal samples and 4 of 6 of the refueling exposure samples, analysis was made using a different packed column gas chromatograph (GC). The parameters for this GC were

1. Column — 12 ft long × 1/8 in. stainless steel column
2. Column packing — 10% SP-2100, 80/100 mesh Supelcoport
3. Carrier gas — nitrogen
4. Carrier gas flow rate — 30 mL/min
5. Temperature program — began at 0°C, then increased at 8°C/min to 154°C, where it was held for 2 min
6. Detector — flame ionization detector

Carbon disulfide was used as the desorbing solvent.

This analysis method detected only hydrocarbons with more than six carbon atoms. The lower limit of detection was 30 μg per sample.

Results

Monitoring Program Data

The monitoring data for the survey of gasoline exposures for truck drivers and terminal operators, marine loading operators, and service station attendants were presented in three ways: (1) arithmetic mean and standard deviation, (2) range, and (3) geometric mean and geometric standard deviation. The data, with the exception of the geometric standard deviation, were presented both as mg/m^3 and ppm. It was apparent to the authors that the exposure data were not normally distributed and that they most likely fit a log-normal or geometric distribution. Consequently, the geometric mean was used to describe exposures where possible.

Large variations in exposure existed for all occupations in this study.

Exposures at Refinery Terminals

The 8-hour TWA exposures for hydrocarbons with more than six carbon atoms at the terminals were on average less than 1.0% of the established American Conference of Governmental Industrial Hygienists Threshold Limit Values (TLV) for gasoline of 900 mg/m^3 (300 ppm).[14] After the exposure, concentrations of hydrocarbons with more than 4 carbon atoms, for the subset of 37 terminal samples, were adjusted to exclude other contributions; they were similar in magnitude to the results for hydrocarbons with more than 6 carbon atoms. Then, no apparent discrepancy existed between the two different GC analytical methods used for the subset of 37 terminal samples and for the remaining 146 terminal samples. The overall geometric mean exposure for the subset of 37 samples was less than 4% of the TLV for gasoline vapors. No statistical difference in exposures among the five terminals was found, regardless of the method of loading or whether or not there was vapor recovery.

Exposures at the Tanker/Barge Loading Facilities

For the marine loading operation, exposure concentrations of hydrocarbons with more than four carbon atoms were approximately three times higher than the comparable exposures at the terminals to hydrocarbons with more than four carbon atoms. However, the marine loading exposures on the average, were only approximately 10% of the TLV. It was, however, not known to what extent the high exposure figure for the marine loading operations was a true reflection of the occupation or if it was due to the different sampling methods (charcoal tubes for the marine loading operation and diffusion badges for the terminal operations). The authors inferred that the marine loading operation exposure measurements were an accurate reflection of the exposures and might even have been low.

Exposures at the Service Plaza

At the service plaza, the 8-hour TWA exposures for hydrocarbons with more than six carbon atoms, were, on the average, 0.4% of the established TLV. Exposures at the service plaza were thus very similar to exposures of truck drivers and operators at the bulk terminals. Large variations of exposure existed as they did at the bulk terminals. The exposures for the service station attendants were four times those which were detectable only following refueling of 25 gal of gasoline and above in the monitoring study conducted to determine exposures to gasoline vapors while refueling cars. For the service plaza survey, two types of monitoring devices were used; the 3M 3500 Organic Vapor Monitor diffusion badge consistently indicated higher exposures.

Conclusions

1. Exposures of consumers to gasoline vapor are anticipated to be significantly lower than the exposures of terminal truck drivers and operators, marine loading operators, and service station attendants. The exposures of these workers are expected to be substantially below the established American Conference of Government Industrial Hygienists Threshold Limit Values for gasoline.

2. n-Butane, isobutane, n-pentane, and isopentane constituted from 61 to 67% by weight of the total gasoline vapor samples.

3. Regardless of the gasoline blend, monitoring (exposure) conditions, or magnitude of exposure, the chemical composition of gasoline vapors appeared to be remarkably constant.
4. Exposures to benzene from gasoline vapors were very low for terminal workers and service station attendants (98% of exposures fell below 1.0 ppm, 8-hour TWA). Exposures of consumers to benzene during automobile refueling were expected to be significantly lower than the above figure.

HIGH-VOLUME SERVICE STATION GASOLINE VAPOR EXPOSURES

Details of a field-tested sampling and analysis technique that provided accurate and reliable ambient air data for gasoline hydrocarbon vapors in the environment of a high-volume service station have been published by Kearney and Dunham.[15]

Because of the maximum gasoline hydrocarbon vapor exposure it provided, a high-volume service station (90,000 gal/month) in Eastern Pennsylvania was chosen as the field site for an exposure survey. The station (surrounded by 10 other gasoline stations within a 1-mile radius) is located on a major interstate highway, 1 mile from a heavily traveled turnpike interchange.

Collection of samples was made for 5 days, Monday through Friday, from 6:00 a.m. to 11:00 p.m. These hours covered three working shifts of personnel of the station and included the morning rush hours and the evening rush hours. It was estimated (from gasoline sales figures tabulated for each hour during the survey) that gasoline purchases were made for 19 cars per hour.

The samples were of the following types:

1. Time weighted average (TWA) for the:
   a. Manager
   b. Captain attendant
   c. Attendant
   d. Mechanic
2. Personal, Short Term for:
   a. Self service (one tank)
   b. Full service (more than one tank, and other services)
   c. Driver/Loader (off-loading)
3. Area:
   a. Pump island — self service and full service
   b. Perimeter delivery area — 20 ft from drop pad

The number of samples varied from 1 to 17, and the average sample time varied from 10 to 445 min.

Sample collection was on 100/50 mg NIOSH-approved charcoal tubes.

To establish the best locations for area sampling (upwind vs. downwind) and to aid in the interpretation of the final data, meteorological information was collected during the sampling period. Throughout the 5 days, wind speed and direction were monitored continuously; wind dispersion was considered to be good. Values of temperature, relative humidity, and barometric pressure were acquired hourly from the Philadelphia International Airport and/or the Willow Grove Air Station, both of which were within 30 miles of the service station. The range of temperature was 40 to 77°F, the range of relative humidity was 21 to 100%, and the range of barometric pressure was 29.75 to 30.48 mm Hg.

Sampling Scheme

A sampling scheme and an analytical method, including standardization techniques, were developed and laboratory tested prior to the field service station evaluation.

The sampling scheme was designed to collect all of the gasoline vapor components and to accomplish the accurate measurement of low concentrations. For the collection and retention of long-term exposure samples of highly volatile low molecular weight gasoline vapor, 100/50 mg charcoal adsorption tubes were used with a flow rate of 100 cm³/min. For short-term exposures, the charcoal tubes were used with a flow rate of 900 cm³/min.

Analysis Procedure

A gas chromatograph (GC) was used for the separate analysis of the vapors desorbed from the front and back sections of the charcoal tubes. The parameters for the gas chromatograph were

1. Column — 60 m fused-silica SE-30 column
2. Detector — flame ionization detector
3. Injection temperature — 150°C
4. Detector setting — 275°C
5. Oven temperature program — held at 0°C for 10 min, programmed from 0 to 100°C at 2°C/min, held at 100°C for 5 min

Decane or dodecane was used as an internal standard. The desorption solvent was methylene chloride.

Standards

For standardization, two techniques were used: the first for the quantitative determination of the total gasoline vapor in each sample, and the second for the quantitative determination and identification of the individual gasoline vapor components in the samples.

Gasoline vapor was generated using the source gasoline. The source gasoline for the service station samples was a mixture of the three grades sold at the station, in ratios which represented the actual sales figures during the sampling period. The vapor mixture was analyzed by GC-mass spectrometry and GC-flame ionization detection under identical conditions. GC retention times were used to identify the individual gasoline vapor components.

For all of the normal hydrocarbons in the gasoline vapor range, specifically from butane to nonane, three-point calibration curves were generated. The use of calibration curves was possible because there was a linear response between GC-flame ionization detection and carbon number.

For the individual components and for the total gasoline hydrocarbon content, an analytical detection limit of 0.01 mg/mL was determined. For some of the paraffins, aromatics, and total gasoline vapor, desorption efficiencies were found to be in the range of 90 to 100%.

Analytical Results

The analytical data showed that the predominant ambient air hydrocarbons were those of C_4 compounds (hydrocarbons with four carbon atoms) and C_5 compounds (hydrocarbons with five carbon atoms).

The results of the monitoring of gasoline vapor exposures showed that the total gasoline Time Weighed Average (TWA) exposures for service station attendants ranged from 0.6 to 4.8 ppm; the geometric mean value was 1.5 ppm. Short-term personal exposure for a driver while off-loading a delivery of gasoline was 7.5 ppm. Short-term personal exposures ranged from undetectable to 38.8 ppm; the geometric mean value was 5.8 ppm.

The exposure levels for individual hydrocarbon components were very low when compared with OSHA standards, ACGIH Threshold Values, or previously published data.

Conclusion

It was concluded that the sampling technique and analytical method, as shown by the data acquired at the service station, was successful for determining levels of gasoline vapor in the ambient in areas of low concentration.

OCCUPATIONAL HYDROCARBON EXPOSURES AT PETROLEUM BULK TERMINALS AND AGENCIES

Verma et al.[8] reported on occupational exposures to 55 hydrocarbon components of gasoline and petroleum products measured at the bulk terminals and agencies of six Ontario, Canada petroleum companies. The methodology of the Ontario Petroleum Association (OPA) was described. Statistical summaries of the concentrations in terms of hydrocarbon components are presented. Exposures of selected jobs and operations are compared. The findings are related to previous work.

The jobs to be sampled that were expected to involve hydrocarbon exposures were (1) terminal truck drivers, (2) agency truck drivers, and (3) bulk terminal plantmen.

The following definitions of terms are useful:

1. Terminal — primary distribution center for petroleum products; it is usually associated with a refinery or pipeline.
2. Agency — secondary distribution centers; petroleum products are distributed from the terminal to the agencies by tanker; petroleum products are distributed from the agencies to gas stations, commercial accounts, and homes.
3. Top-loading — loading petroleum products into a truck through a hatch at the top of the truck.
4. Bottom-loading — pumping petroleum products into a tank using a loading arm which is attached by a sealed connector to the bottom of the truck tank.
5. Off loading — unloading petroleum products from a truck through a hose into a tank.

Sampling

Each of the 6 member companies was to collect 35 (15 long-term and 20 short-term) personal samples and 2 blanks. A sampling sheet was designed for the study. The form was used to record:

1. Air temperature
2. Barometric pressure
3. Relative humidity
4. Wind speed
5. Wind direction
6. Product temperature
7. Types of gas and products loaded during sampling
8. Amount of gas and products loaded during sampling
9. Number of trips taken
10. Whether the worker smoked during the sampling period
11. A company code

Collection was made of representative bulk samples of petroleum products handled by workers being monitored for exposures.

Collection of air samples was on 600-mg activated charcoal tubes. The air was drawn through the tubes for a known time period at a specified flow rate. The flow rate for air sampling was 1 L/min for short-term samples and 50 cm^3/min for long-term samples. The 15 long-term (full-shift) samples were to include 5 truck driver samples each at the bulk terminal and the agency, and 5 plantmen samples at the bulk terminal; 5 top-loading and 5 bottom-loading samples at the bulk terminal, and 5 top-loading and 5 off-loading samples at the agencies comprised the 20 short-term samples. The sampling period for the short-term personal samples taken on truck drivers was at least 10 min and not more than 20 min. Neither the terminals nor agencies in this investigation had vapor recovery systems.

Laboratory Analysis

The bulk samples, unexposed charcoal tubes, exposed charcoal tubes, and corresponding data sheets were shipped for hydrocarbon analysis to McMaster University's (Hamilton, Ontario, Canada) American Industrial Hygiene Association-accredited Occupational Health Laboratory. The samples were analyzed for 55 components of gasoline. The analytical method developed by the Research Triangle Institute and used in an American Petroleum Institute study of exposure to hydrocarbon components of gasoline in the petroleum industry[9] was used, with some minor modifications in the present study to increase calibration accuracy and to achieve lower detection limits. The gas chromatograph used for analysis was calibrated with all of the 55 components prior to analyzing each set of samples.

A gas chromatograph (GC) was used to assess laboratory quality control by the introduction of spiked samples and duplicate analysis. The parameters for the GC were

1. Column — coated 60 m long × 0.32 mm I.D. fused silica capillary column
2. Column coating thickness — 1.0 μm
3. Detector — flame ionization detector

The samples, separated into front and back sections, were desorbed in 2 mL of carbon disulfide. Analysis was by double injections of front sections and a single injection of back sections, using the GC splitless mode. The sampled amount was represented by the mean of the results of two injections.

Conclusions

1. Hydrocarbon concentrations, both long-term and short-term, appeared to be lognormally distributed.
2. Less than 1% of the time, full-shift exposures of bulk terminal drivers, agency drivers, and plantmen to total hydrocarbons (THC) and other measured components for which exposure limits (threshold limit value-time-weighted average, TLV-TWA) existed were found to be above their respective 1986–1987 TLV exposure limits.[14]
3. Less than 3% of the time, short-term exposures of bulk terminal truck drivers for THC and other measured components were found to be above the 1986–1987 TLV exposure limits.[14]
4. About 7 and 17% of the time, short-term exposures of agency truck drivers were found to be above the 1986–1987 TLV short-term exposure limits[14] for THC for top-loading and off-loading, respectively.
5. About 1 and 4% of the time, the short-term exposure of agency truck drivers to benzene during top-loading and off-loading, respectively, was found to exceed the 1986–1987 TLV short-term exposure limits.[14]
6. For long-term exposure to benzene, up to 69% of assessments could be expected to exceed the 1990–1991 proposed TLV-TWA of 0.1 ppm.
7. Full-shift hydrocarbon exposures of agency truck drivers were significantly higher than full-shift exposures of bulk terminal truck drivers.

8. At the bulk terminals with no vapor recovery system, the short-term hydrocarbon exposures during top-loading were significantly higher than those for bottom-loading.

REFERENCES

1. Russo, P. J., G. R. Florky, and D. E. Agopsowicz. "Performance Evaluation of a Gasoline Vapor Sampling Method," *Am. Ind. Hyg. Assoc. J.* 48:528–531 (1987).
2. MacFarland, H. N., C. E. Ulrich, C. E. Holdsworth, D. N. Kitchen, W. H. Halliwell, and S. C. Blum. "A Chronic Inhalation Study With Unleaded Gasoline Vapor," *J. Am. Coll. Toxicol.* 3:231–248 (1984).
3. Weaver, N. K. "Gasoline Toxicology — Implications for Human Health," *Ann. N. Y. Acad. Sci.* 534:441–451 (1988).
4. Halder, C. A., G. S. Van Gorp, N. S. Hatoum, and T. M. Warne. "Gasoline Vapor Exposures. Part I. Characterization of Workplace Exposures," *Am. Ind. Hyg. Assoc. J.* 47:164–172 (1986).
5. Halder, C. A., G. S. Van Gorp, N. S. Hatoum, and T. M. Warne. "Gasoline Vapor Exposures. Part II. Evaluation of the Nephrotoxicity of the Major C_4/C_5 Hydrocarbon Components," *Am. Ind. Hyg. Assoc. J.* 47:173–175 (1986).
6. Wang, O., and G. K. Raabe. "Critical Review of Cancer Epidemiology in Petroleum Industry Employees, with a Quantitative Meta-Analysis of Cancer Sites," *Am. J. Ind. Med.* 15:283–310 (1989).
7. Halder, C. A., T. M. Warne, and N. S. Hatoum. "Renal Toxicity of Gasoline and Related Petroleum Naphthas in Male Rats." *Renal Effects of Petroleum Hydrocarbons. Vol. VII, Advances in Modern Environmental Toxicity,* M. A. Mehlman, G. P. Hemstreet, III, J. J. Thorpe, and N. K. Weaver, Eds. (Princeton, N.J.: Princeton Scientific Publishers, 1984), pp. 73–88.
8. Verma, D. K., J. A. Julian, G. Bebee, W. K. Cheng, K. Holburn, and L. Shaw. "Hydrocarbon Exposures at Petroleum Bulk Terminals and Agencies," *Am. Ind. Hyg. Assoc. J.* 53:645–656 (1992).
9. Rappaport, S. M., S. Selvin, and M. A. Waters. "Exposure to Hydrocarbon Components of Gasoline in the Petroleum Industry," *Appl. Ind. Hyg.* 2:148–154 (1987).
10. National Institute for Occupational Safety and Health. *NIOSH Manual of Analytical Methods: New Publications* (Cincinnati, Ohio: NIOSH, 1977) p. 127.
11. National Institute for Occupational Safety and Health. *Development and Validation of Methods for Sampling and Analysis of Workplace Toxic Substances* (Cincinnati, Ohio: NIOSH, 1979) p. 13.
12. Melcher, R., R. R. Langner, and R. O. Kagel. "Criteria for Evaluation of Methods for the Collection of Organic Pollutants in Air Using Solid Sorbents," *Am. Ind. Hyg. Assoc. J.* 39:349 (1978).
13. Ward, W. L. Personal Communication.
14. American Conference of Governmental Industrial Hygienists. *TLVs [TM]-Threshold Limit Values for Chemical Substances and Physical Agents in the Work Environment with Intended Changes for 1983–1984* (Cincinnati, Ohio: ACGIH, 1983) p. 21.
15. Kearney, C. A., and D. B. Dunham. "Gasoline Vapor Exposures at a High Volume Service Station," *Am. Ind. Hyg. Assoc. J.* 47: 535–539 (1986).

CHAPTER 8

Chemical Warfare Agents

INTRODUCTION

Methods that have been used to detect and identify chemical warfare agents include the following:

1. Capillary column gas chromatography-flame ionization detection for routine screening of samples for presence of chemical warfare agents[1,2]
2. Mass spectrometry for identification of agents and their degradation products[3]
3. Electron impact ionization for mass spectrometry for verification of organophosphorus chemical warfare agents[4-9]
4. Tandem mass spectrometry (MS-MS) for confirmation of trace levels of organophosphorus chemical warfare agents[10]
5. Capillary column ammonia chemical ionization gas chromatography-mass spectrometry and gas chromatography-tandem mass spectrometry for detection and confirmation of sarin and soman in a complex airborne matrix[3]
6. Miniature solid adsorbent sampling tubes and thermal desorption gas chromatography system[11]
7. Color or precipitation spot test paper strips or gas test tubes[12]
8. Gas liquid chromatography-mass spectrometry with electron impact ionization[13]
9. Gas chromatography with electron capture detection[14]
10. Gas chromatography with multidetector finish[14]
11. Fused silica capillary column gas chromatography with columns coated by DB-1, DB-5, DB-1701, and DBWAX[15]

12. Chemical ionization mass spectrometry[16,17]
13. Thin-layer chromatography[18,19]
14. High-performance liquid chromatography[20]
15. Solid sorbent sampling-gas chromatography[21]

EXPERIMENTAL DETERMINATIONS OF CHEMICAL WARFARE AGENTS

Amine-Coated Solid Adsorbent-Gas Chromatography Determination of Phosgene and Chloroformates

Hendershott[21] described a method for the determination of phosgene and a number of reactive chloroformates in air at low levels of concentration. Air containing these contaminants was collected on di-*n*-butylamine-coated solid adsorbent. Carbamate and urea derivatives formed were desorbed with hexane, washed with 1 *N* hydrochloric acid, and analyzed by gas chromatography with flame ionization detection (FID).

The reactant-coated adsorbent material was packed (an average of 0.36 ± 0.02 g in a single 3.5-cm section) into standard 600-mg sampling tubes. The adsorbent was retained in the tubes by glass wool plugs with steel retaining wires. The ends of the tubes were fire-sealed.

Portable personal sampling pumps were used in field sampling using the amine-coated solid adsorbent tubes. A total volume of 24 L per tube was collected during 8 hours at a sampling rate of 50 cm^3/min. Desorptions of the collected samples were carried out at room temperature.

Two sets of instrument parameters were used for the gas chromatographic analysis to yield either phosgene alone or phosgene together with up to eight chloroformates in a single chromatographic analysis. A gas chromatograph with a flame ionization detector (FID) was used.

The parameters for the gas chromatograph were

1. Column — 3 m × 1/8 in. O.D. nickel SP-Alloy
2. Column packing — 100/120 mesh Supelcoport coated with 10% neopentyl glycol succinate (NPGS)
3. Carrier gas — helium
4. Carrier gas flow rate — 24 mL/min
5. Column temperature programs — (a) isothermal at 190°C, (b) initial temperature of 150°C held for 16 min, increased at 16°C/min to 190°C, held at 190°C for 16 min
6. Injection port temperature — 200°C

The average recovery efficiency ranged from 98 to 106%, with an average detection limit of 0.7 μg or 0.08 ppm (volume/volume) in 1.5 L of air. The method provided the most promising technique developed at that time for determining phosgene, several chloroformates, and carboxylic acid chlorides in air.

GC-MS Study of Munitions Grade Tabun Containing Impurities

Combined capillary column gas chromatography-mass spectrometry (GC-MS), under both electron impact (EI) and chemical ionization (CI) conditions, was used to study a sample of munitions grade tabun known to contain several impurities.[6] Based on their mass spectral data, five impurities (three unreported previously in tabun) were identified and characterized.

A gas chromatograph equipped with a flame ionization detector (FID) was used for all capillary column GC-FID analyses. A double focusing mass spectrometer was used to perform capillary column GC-MS analyses. The operating parameters for the gas chromatograph were

1. Columns — three 15 m × 0.32 mm I.D. capillary columns
2. Column coatings — 0.25 μm thick:
   a. DB-1 (100% dimethylpolysiloxane)
   b. DB-5 (95% methyl-(5%)-diphenylpolysiloxane)
   c. DB-1701 (86% dimethyl-(14%)-cyanopropylphenylpolysiloxane)
3. Injections — on-column, using an injector designed by the authors
4. Injection temperature — 50°C
5. Temperature program — 50°C for 2 min, followed by 10°C/min to 300°C
6. Carrier gas—high-purity helium
7. Carrier gas linear velocity — 35 cm/sec (methane injection at 50°C)

The GC and mass spectral data were sufficient to identify tabun and its impurities. Ammonia CI was particularly useful.

Gas Chromatographic Retention Indices of Chemical Warfare Agents and Simulants

Fused silica capillary columns coated with DB-1, DB-5, DB-1701, and DBWAX films were used to determine retention indices, relative to a homologous n-alkane series, for 22 chemical warfare agents and simulants.[1] Organophosphorus chemical warfare agents, vesicants, irritants, and various simulants were included in the study.

Retention time can be defined[22] as "the elapsed time from injection of the sample to recording of the peak maximum of a component band." D'Agostino and Provost[1] calculated the retention index using the equation:

$$RI_c = 100n[(t_c - t_z)/(t_{(z+n)} - t_{(z)})] + 100z \qquad (1)$$

where RI_c is the retention index; n is the difference in carbon number between two n-alkanes on either side of the compound, c, of interest; t is the retention time; and z is the carbon number of the n-alkane immediately prior to the compound c.

The 22 compounds for which the retention indices were determined are

A. Organophosphorus Compounds
1. Sarin (isopropyl methylphosphonofluoridate)
2. Soman (pinacolyl methylphosphonofluoridate)
3. Tabun (ethyl *N,N*- dimethylphosphoroamidocyanidate)
4. 2-Methylcyclohexyl methylphosphonofluoridate
5. *o*-Ethyl S-2-diisopropylaminoethyl methylphosphonothiolate
6. Triethyl phosphate
7. Dimethyl morpholinophosphoramidate
8. Tributyl phosphate

B. Vesicants and Related Compounds
9. 1,4-Dithiane
10. Mustard (bis(2-chloroethyl)sulfide)
11. Thiodiglycol (2,2'-thiodiethanol)
12. Bis(2-chloroethyl)disulfide
13. Bis(2-chloroethyl)trisulfide
14. Sesquimustard
15. 1,1'-Oxybis[2-((2-chloroethyl)thio)ethane]

C. Irritants
16. 1-Methyoxycycloheptatriene
17. CN (2-chloroacetophenone)
18. CS (*o*-chlorobenzylidenemalononitrile)
19. Dibenz[*b,f*]-1,4-oxazepin

D.Simulants
20. DMSO (dimethyl sulfoxide)
21. $DMSO_2$ (dimethyl sulfone)
22. Methyl salicylate

A gas chromatograph with a flame ionization detector (FID) was used for all analyses. The operating parameters for the gas chromatograph were

1. Columns — four bonded and cross-linked 15 m × 0.32 mm I.D. capillary columns
2. Column coatings — DB-1, DB-5, DB-1701, or 100% polyethylene glycol films
3. Coating film thickness — 0.25 μm
4. Injections — on-column at 50°C
5. Temperature program — 50°C for 2 min and then 10°C/min to 300°C for the DB-1, DB-5, and DB-1701 columns and to 250°C for the polyethylene glycol-coated column; the upper temperature was maintained for 5 min
6. Carrier gas — high-purity helium
7. Carrier gas linear velocity — 35 cm/sec (methane injection at 50°C)

The reproducibility of the retention indices calculated was excellent. A soil sample containing chemical warfare agents was analyzed to demonstrate the applicability of the method for compound verification.

DETECTION OF SARIN, SOMAN, AND MUSTARD FROM A DIESEL EXHAUST ENVIRONMENT

Full scanning gas chromatography-mass spectrometry (GC-MS) was used by D'Agostino et al.[10] to detect and confirm the chemical warfare agents sarin, soman, and mustard at the nanogram level in spiked extracts of a diesel exhaust environment sampled onto the charcoal of a Canadian C2 respirator canister. This airborne diesel exhaust matrix was used to develop a GC-MS-MS approach for the identification of sarin, soman, and mustard.

Diesel exhaust environment air was sampled through the charcoal canister, at the typical working respiratory rate of 20 L/min, for 4 hours. The 108 g of charcoal in the canister was Soxhlet extracted with 250 mL of dichloromethane for 6 hours. The extract was concentrated to 10 mL under a gentle stream of nitrogen. Spikes of 50 µg/mL of sarin, 5 µg/mL of soman, and 500 ng/mL of mustard were added to the extract.

All GC analyses of spiked and unspiked charcoal extracts were made on injection volumes of 0.4 and 1 µL, equivalent to 0.0002 and 0.0005 m³, respectively, of air sampled on the charcoal of the C2 canisters. Capillary column GC-flame ionization detector (FID) analyses were performed on a gas chromatograph equipped with an on-column injector designed by the authors.

The operating parameters for the GC were

1. Column — 15 m × 0.32 I.D. capillary column
2. Column coating — 0.25 µm DB-5 film
3. Temperature program — 40°C for 2 min, then 10°C/min to 280°C for 5 min

Capillary column GC-MS analyses were performed with a double focusing mass spectrometer interfaced with a gas chromatograph under chromatographic conditions identical to those for the GC-FID analyses.

Capillary column GC-MS-MS analyses were performed with a hybrid tandem mass spectrometer equipped with a gas chromatograph. The operating parameters were

1. Column — 15 m × 0.32 mm I.D. capillary column
2. Column coating — DB-5 film
3. Temperature program — 40 or 50°C for 2 min, then 10°C/min to a maximum of 280°C

Electron impact (EI)-GC conditions were identical to those for the GC-MS analyses, except that the source temperature was 250°C rather than 200°C and the accelerating voltage was 8 kV rather than 6 kV.

Sarin, soman, and mustard were detected and confirmed during capillary column GC-EI-MS conditions at nanogram levels in spiked extracts of diesel exhaust environment. The complexity of the sample extract made GC-FID of little utility for this application. Chemical interferences associated with the samples were virtually eliminated. Low-picogram GC-MS-MS detection limits were estimated for sarin, soman, and mustard in the presence of numerous interfering components in diesel exhaust and in the charcoal bed. The diesel exhaust environment sampled was similar in composition to the volatile battle-field components extracted from a respirator canister circulated as part of an interlaboratory analytical exercise.

GC-MS-MS was found to be the most sensitive for the confirmation of sarin, soman, and mustard in the presence of components commonly found in an airborne battlefield environment.

SARIN AND SOMAN DETECTION IN A COMPLEX AIRBORNE MATRIX

Full scanning capillary column ammonia chemical ionization mass spectrometry was used to detect and confirm the chemical warfare agents sarin and soman at nanogram levels in spiked extracts of a diesel exhaust environment sampled onto the charcoal of a Canadian C2 respirator canister.[3] Evaluation of capillary column ammonia chemical ionization tandem mass spectrometry as a possible verification technology was made in the diesel exhaust environment, which is typical of battlefield conditions.

To develop and evaluate ammonia chemical ionization-mass spectrometry (CI-MS) and ammonia CI-MS-MS for the detection and confirmation of sarin and soman in a complex airborne matrix, a capillary column gas chromatography (GC) study was made.

Diesel exhaust environment air was sampled through the charcoal canister, at the typical working respiratory rate of 20 mL/min, for 4 hours. The charcoal, 108 g, was Soxhlet extracted with 250 mL of dichloromethane for 6 hours. The extract was concentrated to 10 mL under a gentle stream of nitrogen. The extracts were spiked at several levels of sarin and soman.

A double focusing mass spectrometer interfaced to a gas chromatograph was used for the capillary column ammonia chemical ionization GC-MS full scanning and selected ion monitoring analyses. Injections were on-column at 40°C. The operating parameters for the GC were

1. Column — 15 m × 0.32 mm I.D. capillary column
2. Column coating — 0.25 μm DB-5 film
3. Temperature program — 40°C for 2 min, then 10°C/min to 280°C.

A hybrid tandem mass spectrometer equipped with a gas chromatograph was used for the capillary column ammonia chemical ionization analyses.

Injections were on-column at 40°C. The operating parameters for the gas chromatograph were

1. Column — 15 m × 0.32 capillary column
2. Column coating — 0.25 μm DB-5 film
3. Temperature program — 40°C for 2 min, then 8°C/min or 20°C/min to a maximum of 280°C

The source conditions for the ammonia CI were identical to those for the GC-MS analysis except that the accelerating voltage was 8 kV rather than 6 kV and the emission was 1000 μA rather than 500 μA.

The detection limits were 40 pg and just above 500 pg for sarin and soman, respectively, in the spiked extracts for selected ion monitoring. For capillary column ammonia CI-MS-MS analysis of these agents, chemical interferences were significantly reduced and detection limits were 15 and 80 pg, respectively, for sarin and soman in the presence of matrix component concentrations of 2 or 3 orders of magnitude greater than the spiked agents.

The detection of organophosphorus chemical warfare agents by the application of tandem MS, under ammonia CI conditions, "appears to be an attractive approach for the verification of 'target' compounds in complex environmental matrices such as those that may be encountered during the airborne sampling of battlefield emissions."[3]

IDENTIFICATION OF TABUN IMPURITIES

Combined capillary column gas chromatography-mass spectrometry (GC-MS) under electron impact (EI) and ammonia and deuterated ammonia chemical ionization (CI) conditions were used to study a munitions-grade tabun sample containing a number of sample components.[9] A total of 19 impurities were identified.

A gas chromatograph with flame ionization detection (FID) was used for all capillary column GC-FID analyses. The operating parameters for the gas chromatograph were

1. Column — 15 m × 0.32 mm I.D. capillary column
2. Column coating — DB-5, DB-1701, and DBWAX films
3. Injections — on-column injections at 50°C
4. Temperature program — 50°C for 2 min then 10°C/min to 260°C, 280°C, and 230°C for 5 min for DB-5, DB-1701, and DBWAX columns, respectively
5. Carrier gas — high-purity helium
6. Linear velocity of carrier gas — 35 cm/sec (methane injection at 50°C)

A double focusing mass spectrometer interfaced with a gas chromatograph was used for the capillary column GC-MS analyses. Capillary column GC-MS under EI, ammonia CI, and deuterated ammonia CI conditions was used to analyze samples of tabun. Impurities accounted for 28% of the volatile organic content of a munitions-grade sample, as indicated by capillary column GC-FID analysis.

The mass spectral data provided by the study were sufficient for identification of tabun and related impurities. Ammonia and deuterated ammonia CI provided molecular ion information for all the identified sample components, and was thus particularly useful.

SOLID ADSORBENT SAMPLING AND ANALYSIS SYSTEM

Hancock et al.[11] described the design, operation, and initial chromatographic performance of an integrated solid adsorbent sampling and analysis system for organic compounds in air with application to compounds of chemical defense interest. Integration of field sampling and laboratory analysis through a common sampling component was the principal objective of the system. The system is known as the Minitube Air Sampling System (MASS) and was designed at the Canadian Defence Research Establishment Suffield (Ralston, Alberta).

Borosilicate glass minitubes, 38 mm × 2 mm I.D., were packed with approximately 14 mg of Tenax TA (Chrompack) and loaded into an air sampler for sample collection. After sample collection, the minitube holder was taken to the laboratory and inserted in an automatic thermal desorption (TD) system for gas chromatographic (GC) analysis. Two gas chromatographs were used — one equipped with a flame ionization detector (FID) and the other equipped with a flame photometric detector (FPD).

The operating parameters for the gas chromatograph systems were

1. Columns — 5 m × 0.53 mm fused silica capillary columns
2. Column coatings — either a DB-5 or a DB-1701 film
3. Carrier gas — helium
4. Carrier gas flow rate — 5 to 20 mL/min
5. Typical temperature program — 50°C for 2 min, then 10°C/min to 230°C

Standard solutions were of soman (GD), mustard (HD), and of the simulants methyl salicylate (MS) and triethyl phosphate (TEP). The standard solutions were prepared in glass-distilled acetone and loaded onto minitubes.

For field evaluation of the system, air above a vessel suspected of containing a chemical warfare agent was sampled by drawing 50 mL of the vapor through a minitube packed with Tenax TA. The sorbent was held in place in minitubes with silanized glass wool or stainless steel screens.

The minitube air sampling system was the first system capable of the automatic collection of samples and unattended chromatographic analysis. Chromatographic separations had a precision of less than 5% in peak area, and less than 1% in retention time. The system was field tested and used to successfully collect and analyze an unknown vapor sample containing the chemical warfare agent mustard.

MULTIDETECTOR TECHNIQUE FOR GAS CHROMATOGRAPHIC ESTIMATION OF SOME SULFUR AND NITROGEN MUSTARDS

A simultaneous multidetector system for gas chromatography was applied to a series of sulfur and nitrogen mustard compounds by Sass and Steger.[14]

The system consisted of a gas chromatograph which could hold four different columns, each of which could be connected to a separate detector or the ouput from any single column could be split two or three ways for separate and simultaneous detection. The detection could be made by an electron capture detector (ECD) (pulsed or d.c. mode), a flame ionization detector (FID), and a flame photometric detector (FPD), or by a three-way combination that could include the Coulson or the Hall nitrogen detector. Phosphorus and sulfur were detected simultaneously on a dual-head flame photometer.

The chromatograph peaks could be quantitated by one of several integrators. Chromatography was performed both under temperature-programmed conditions and isothermally for aiding in the identification of the species and for determining the most reasonable isothermal temperatures for analysis of specific compounds.

Operating gas chromatograph parameters common to all mustards and parathion were

1. Column — 6 ft × 2 mm I.D. Pyrex glass
2. Column packing — 3% QF-1 on 100/200 mesh Gas-Chrom Q
3. Column temperature program — 8°C/min from 60 to 280°C; also isothermal at various temperatures
4. Inlet temperature — 200°C
5. Transfer temperature at valves — 210°C
6. Detector temperatures
   a. FID — 250°C
   b. FPD — 185°C
   c. d.c. ECD — 220°C
   d. pulsed mode ECD — 250°C
   e. Hall electrolytic conductivity detector solvent transfer — 220°C
   f. Furnace — 875°C
7. Carrier gas
   a. helium at 40 mL/min for all detectors except:
      a1. argon-methane at 90 mL/min for pulsed mode ECD
      a2. preferably nitrogen at 30 mL/min for d.c. ECD only when it is not in a multidetector phase

Up to four detectors could be operated simultaneously with a single injection of nanograms to micrograms of sample. Parathion was used since it was sensitive to all the detectors. The Sass and Steger paper[14] described their application of the multidetector concept to the determination of trace quantities of some mustards, and parathion as a representative of a phosphorus pesticide.

GAS CHROMATOGRAPHIC RETENTION INDICES FOR SULFUR VESICANTS AND RELATED COMPOUNDS

D'Agostino and Provost[2] determined temperature-programmed retention indices, relative to an *n*-alkane homologous series, for 37 sulfur vesicant or vesicant-related compounds using films of DB-1, DB-5, and DB-1701 on fused silica capillary columns. The compounds were identified during the analysis of munitions-grade mustard, an Iran/Iraq soil sample, and aqueous samples containing hydrolyzed mustard.

A gas chromatograph equipped with a flame ionization detector (FID) was used for all analyses. Operating parameters for the gas chromatograph were

1. Columns — 3 bonded and cross-linked 15 m × 0.32 I.D. capillary columns, coated with 0.25 µm films:
   a. Column (a) film — DB-1
   b. Column (b) film — DB-5
   c. Column (c) film — DB-1701
2. Injection — on-column
3. Injection temperature — 50°C
4. Temperature program — 50°C for 2 min then 10°C/min to 300°C, which was maintained for 5 min
5. Carrier gas — high-purity helium
6. Carrier gas linear velocity — 35 cm/sec (methane injection at 50°C)

The temperature-programmed retention indices should be useful for the verification of sulfur vesicants, their decomposition products, and related impurities in mustard-containing samples.

CAPILLARY COLUMN GAS CHROMATOGRAPHY-MASS SPECTROMETRY ANALYSIS OF VX

D'Agostino et al.[8] studied a sample of O-ethyl S-[2-(diisopropylamino)ethyl] methylphosphonothiolate (VX), known to contain impurities, by combined capillary column gas chromatography-mass spectrometry under both electron impact and chemical ionization conditions. The principal objective of the study was the identification of the VX impurities. Chemical ionization-mass spectrometry (CI-MS) was used to provide molecular ion information required for identification purposes.

The sample of VX had been stored for 10 to 15 years in a glass container. Aliquots of the VX were diluted with HPLC-grade chloroform and stored in PFTE-lined screw-capped glass vials prior to gas chromatographic analyses.

Operating parameters for the gas chromatograph were

1. Columns — 15 m × 0.32 mm I.D. capillary columns
2. Column coatings — 0.25 µm films of DB-1, DB-5, and DB-1701
3. Detector — flame ionization detector (FID)

4. Injector — on-column, of the authors' design
5. Injection temperature — 50°C
6. Temperature program — 50°C for 2 min, then 10°C/min to a maximum of 280°C, which was maintained for 10 min
7. Carrier gas — high-purity helium
8. Carrier gas linear velocity — 35 cm/sec (methane injection at 50°C)

Ammonia chemical ionization was found to be an excellent technique for the detection and identification of many previously unreported impurities in VX. The gas chromatographic and mass spectral data were sufficient for the future identification of VX and a number of VX-related impurities.

More than 20 impurities in the sample of VX, many of which had been been previously reported, were identified and characterized.

REFERENCES

1. D'Agostino, P. A., and L. R. Provost. "Gas Chromatographic Retention Indices of Chemical Warfare Agents and Simulants," *J. Chromatogr.* 331:47 (1985).
2. D'Agostino, P. A., and L. R. Provost. "Gas Chromatographic Retention Indices of Sulfur Vesicants and Related Compounds," *J. Chromatogr.* 436:399 (1988).
3. D'Agostino, P. A., L. R. Provost, and P. W. Brooks. "Detection of Sarin and Soman in a Complex Airborne Matrix by Capillary Column Ammonia Chemical Ionization Gas Chromatography-Mass Spectrometry and Gas Chromatography-Tandem Mass Spectrometry," *J. Chromatogr.* 541:121 (1991).
4. *Chemical and Instrumental Verification of Organophosphorus Warfare Agents,* The Ministry of Foreign Affairs of Finland, Helsinki, 1977.
5. Sass, S., and T. L. Fisher. *Org. Mass Spectrom.* 14:257 (1979).
6. D'Agostino, P. A., A. S. Hansen, P. A. Lockwood, and L. R. Provost. "Capillary Column Gas Chromatography-Mass Spectrometry of Tabun," *J. Chromatogr.* 347:257 (1985).
7. Wils, E. R. J., and A. G. Hulst. *Org. Mass. Spectrom.,* 21: 763 (1986).
8. D'Agostino, P. A., L. R. Provost, and J. Visentini. "Analysis of O-Ethyl S-[2-(Diisopropylamino)Ethyl] Methylphosphonothiolate (VX) by Capillary Column Gas Chromatography-Mass Spectrometry," *J. Chromatogr.* 402:221 (1987).
9. D'Agostino, P. A., L. R. Provost, and K. M. Looye. "Identification of Tabun Impurities by Combined Capillary Column Gas Chromatography-Mass Spectrometry," *J. Chromatogr.* 465:271 (1989).
10. D'Agostino, P. A., L. R. Provost, J. F. Anacleto, and P. W. Brooks. "Capillary Column Gas Chromatography-Mass Spectrometry and Gas Chromatography-Tandem Mass Spectrometry Detection of Chemical Warfare Agents in a Complex Airborne Matrix," *J. Chromatogr.* 504:259 (1990).
11. Hancock, J. R., J. M. McAndless, and R. P. Hicken. "A Solid Adsorbent Based System for the Sampling and Analysis of Organic Compounds in Air: An Application to Compounds of Chemical Defense Interest," *J. Chromatogr. Sci.* 29:40–45 (1991).

12. Klimmek, R., L. Szinicz, and N. Weger. *Chemische Gifte und Kampfstoffe Wirkung und Therapie,* (Stuttgart: Hippokrates Verlag, 1983), p. 57.

13. Vycudilik, W. "Detection of Mustard Gas Bis (2-Chloroethyl)-Sulfide in Urine," *Forensic Sci. Int.* 28:131–136 (1985).

14. Sass, S., and R. J. Steger. "Gas Chromatographic Differentiation and Estimation of Some Sulfur and Nitrogen Mustards Using a Multidetector Technique," *J. Chromatogr.* 238:121–132 (1982).

15. Ali-Mattila, E., K. Siivinen, H. Kenttamaa, and P. Savolahti. *Int. J. Mass Spectrom. Ion Phys.* 47:371 (1983).

16. Wils, E. R. J., and A. G. Hulst. *Fresenius' Z. Anal. Chem.* 321:471–474 (1985).

17. D'Agostino, P. A., and L. R. Provost, *Proceedings of the 35th Annual Conference on Mass Spectrometry and Allied Topics, Denver, CO, May 24–29, 1987,* pp. 147–148.

18. Sass, S., and M. H. Stutz. *J. Chromatogr.* 213:173 (1981).

19. Appler, B., and K. Christmann. *J. Chromatogr.* 264:445 (1983).

20. Bossle, P. C., J. J. Martin, E. W. Sarver, and H. Z. Sommer. *J. Chromatogr.* 283:412 (1984).

21. Hendershott, J. P. "The Simultaneous Determination of Chloroformates and Phosgene at Low Concentrations in Air Using a Solid Sorbent Sampling-Gas Chromatographic Procedure," *Am. Ind. Hyg. Assoc. J.* 47:742 (1986).

22. Ettre, L. S., and A. Zlatkis. *The Practice of Gas Chromatography* (New York: Interscience Publishers, 1967), p. 37.

CHAPTER 9

Polycyclic Aromatic Hydrocarbons (PAHs)

INTRODUCTION

Polycyclic aromatic hydrocarbons (PAHs) result from incomplete combustion of organic (carbonaceous) material[1] and are often associated with industrial processes.[2] They are considered to be the most toxic class of contaminants in emissions from these processes.[2] Many of the PAHs are known carcinogens,[3] and cause increasing concern in the ecosystem.

The formation of PAHs is due to two major causes:[3] (1) endogenic synthesis in the environment by microorganisms, phytoplankton, algae, and highly developed plants, which create the natural background, and (2) man-controlled high-temperature pyrolithic reactions, open burning, and natural volcanic activities.

Man-controlled and/or man-induced combustion processes are quantitatively the most significant, by far. Many of the sources are various industrial processes and various forms of combustion:[3]

1. Carbon black processes
2. General chemical processes
3. Catalytic cracking in the petroleum industry
4. Coal tar pitch processes
5. Asphalt hot road mix processes
6. Coke production in the iron and steel industries
7. Heating and power generation (coal, oil, gas, wood)

8. Commercial enclosed incineration
9. Industrial enclosed incineration
10. Agricultural open fires
11. Coal refuse open fires
12. Forest open fires
13. Combustion in internal combustion engines:
    (a) automobiles
    (b) trucks and buses
14. Volcanic activity

Two forms of degradation of PAHs exist: physical oxidation and biological reduction.

Both in the atmosphere and in the water environment, the most important process is photooxidation.[3]

EXPERIMENTAL DETERMINATIONS OF PAHs

Sampling System for Collection of PAHs

Thrane and Mikalsen[1] described a sampling system for the collection of polycyclic aromatic hydrocarbons. In their investigation, they selected from the literature a sampling system suitable for high-volume sampling of gaseous and particulate PAHs in ambient air and examined its efficiency. The system was to be used for the determination of PAHs levels in background areas and areas close to the sources, with a collection time that did not exceed 24 hours.

In the selected system, PAHs were collected on glass fibers and plugs of polyurethane foam by high-volume sampling. Volatile compounds were trapped on two cylindrical plugs of polyurethane foam located behind a filter. The plugs were 5 cm thick with a diameter of 11 cm and a density of 25 kg/m^3. A vacuum pump was used for sampling.

Samples were collected at a background station located on grass a few hundred meters from an unoccupied farm, in a suburban area, and at three urban sites. Extraction from plugs and filters was made with purified cyclohexane in a Soxhlet apparatus. Gas chromatographic analyses were made on a gas chromatograph fitted with a glass capillary column. The operating parameters for the gas chromatograph were

1. Column — 35 m × 0.34 mm glass column
2. Column coating — SE54
3. Detector — flame ionization (FID)
4. Injection — splitless
5. Temperature of injection/detection block — 275°C
6. Oven temperature program — 100 to 250°C, at 3°C/min
7. Carrier gas — helium
8. Carrier gas flow rate — 4 mL/min

The PAHs determined at three different areas (at all three areas unless otherwise indicated) were

1. Naphthalene
2. Biphenyl
3. Fluorene
4. Dibenzothiophene
5. Phenanthrene
6. Anthracene
7. Carbozole
8. 2-Methylanthracene (urban only)
9. 1-Methylphenanthrene (urban only)
10. Fluoranthene
11. Pyrene
12. Benzo[a]fluorene
13. Benzo[b]fluorene
14. Benzo[a]anthracene
15. Chrysene/triphenylene
16. Benzo[b]fluoranthene (urban only)
17. Benzo[e]pyrene
18. Benzo[a]pyrene
19. Perylene
20. o-Phenylenepyrine
21. Dibenzo[ac]/[ah] anthracene (suburban and urban)
22. Benzo [g,h,i] perylene
23. Coronene (suburban and urban)

The polyurethane foam seemed to be efficient for the collection of selected polycyclic aromatic hydrocarbons. The concentrations of PAHs for which the system was used ranged from 20 to about 1500 ng/m^3. Profiles of PAHs species obtained at different locations were similar. Distributions of gas-phase PAHs on the polyurethane corresponded well with results for a Belgian study of gas-aerosol equilibrium.

Sampling and Analytical Method for PAHs

The physical and chemical nature of the exposure of workers to polycyclic aromatic hydrocarbons were studied by Lesage et al.,[2] using a sampling and analytical method. A gravimetric method was used for an estimation of organic contaminants in order to use a gas chromatograph-mass spectrometer (GC-MS) to analyze PAHs. The GC-MS was also used to sample gaseous and particulate PAHs to determine the physical nature of the exposure.

A glass fiber filter was used to capture particulate contaminants. This was followed by a Chromosorb 120 tube to hold back the gaseous fraction. A sampled air volume of 960 L and a pump flow rate of 2.0 L/min were used. The NIOSH method for benzene-soluble materials,[4] as modified for aluminum refineries,[5] was used. In the laboratory, at a sampling rate of 2 L/min for 8 hours, the observed limit of detection was 50 μg/m^3 with a precision level of 50% within a 95% confidence interval.

The operating conditions for the GC-MS used to analyze the PAHs were

1. Injection — splitless
2. Injector temperature — 280°C
3. Separation column — a DB-5 30 m × 0.25 mm capillary column
4. Column coating film thickness — 0.25 μm
5. Oven temperature program — 60°C for 1 min, then at 10°C/min up to 280°C, and finally at 280°C for 20 min
6. Carrier gas — helium
7. Carrier gas flow rate — 1 mL/min

The GC-MS analysis was done on both the gaseous fraction collected on the sorbent tubes and on benzene-soluble materials. The GC-MS had a precision level of 10% within a 95% confidence interval. For a sample of 960 L, concentrations as low as 0.01 μg/m^3 could be detected.

Extensive field testing of the method was done. The results were affected by sampling temperature, and organic and inorganic interferences. An evaluation of worker exposure to PAHs was obtained using a combination of the gravimetric method and the particulate and gaseous concentration profile of 12 PAHs. The 12 PAHs were

1. Naphthalene	7. Fluoranthene
2. Biphenyl	8. Pyrene
3. Acenaphthene	9. Chrysene
4. Fluorene	10. Benzo[a]anthracene
5. Phenanthrene	11. Benzo[a]pyrene
6. Anthracene	12. Benzo[e]pyrene

If sampling conditions and inteferences by oil and dust were taken into consideration, Lesage et al.[2] considered that the environmental evaluation protocol could be used to monitor and model worker exposure to PAHs.

PAHs and Nitroarene Concentrations in Ambient Air

Arey et al.,[6] in order to measure the concentrations of certain nitroarenes and their parent species in ambient air during a wintertime high-NO$_x$ episode in California, collected ambient air samples on three media: (1) high-volume filters, (2) polyurethane foam (PUF) plugs, and (3) Tenax GC solid adsorbent.

Ambient air was sampled from a roof (elevation about 9 m) at El Camino Community College in Torrance, CA on February 24–25, 1986 during two 12-hour sampling periods, one nighttime (6:00 p.m.–6:00 a.m.) and one daytime (6:00 a.m.–6:00 p.m.). The maximum daytime temperature was 35°C, at 11:00 a.m., and the nighttime temperature was approximately constant at 20°C. Maximum pollutant levels were

1. NO_x — greater than 500 ppb at 8:00 a.m.
2. NO — 400 ppb at 8:00 a.m.
3. NO_2 — 250 ppb at 10:00 a.m.
4. O_3 — 90 ppb at 4:00 p.m.

Tenax GC cartridges were used to sample gas-phase PAHs. Pyrex tubes, 10 cm × 4 mm I.D., were packed with 0.1 g of Tenax GC and doped with 610 ng of naphthalene-d_8. A 1.15-L/min flow rate through the cartridges yielded an approximate sample volume of about 0.8 m^3 for each 12-hour sampling period. To check for breakthrough from the first cartridge, a second Tenax GC cartridge was placed in series downstream — 3 mL of diethyl ether was used to elute the cartridges. Internal standards phenanthrene-d_{10} (60 ng) and anthracene-d_{10} (67 ng) were added to the eluate. The eluate was then concentrated using a micro Snyder apparatus. Gas chromatography-mass spectrometry (GC-MS) with multiple ion detection (MID) was used to analyze the concentrated extracts.

Air samples were collected from 610 m^3 of air using a single high-volume sampler system consisting of a Teflon-impregnated glass fiber filter followed by three polyurethane foam plugs. The plugs were about 9 cm in diameter and 5 cm thick. The system was run at a standard volume flow rate of 30 standard ft^3/min for 12-hour intervals. Ultimately, PAHs and nitroarene fractions were analyzed by GC-MS-MID.

Up to four Teflon-impregnated glass fiber filters from simultaneous high-volume collections of particulate organic matter were combined in order to collect particulate organic matter for analyses of nitrofluoranthenes and nitropyrenes. Total volumes of air were 1960 m^3 from 6:00 a.m. to 6:00 p.m., and 3190 m^3 from 6:00 a.m. to 6:00 p.m.. Ultimately, nitrofluoranthene and nitropyrene fractions were analyzed by GC-MS-MID.

A quadrupole GC-MS operating in the electron impact mode was used for compound identifications and quantifications. The GC column was a 30-m DB-5 fused silica capillary column. The injection system was a cool on-column system. The GC eluted directly into the MS ion source.

The PAHs and nitro-derivatives measured were

PAHs
1. Naphthalene
2. 2-Methylnaphthalene
3. 1-Methylnaphthalene
4. Biphenyl
5. Phenanthrene
6. Anthracene
7. Fluoranthene
8. Pyrene
9. Benzo[e]pyrene
10. Benzo[a]pyrene
11. Perylene

Nitroarenes
1. 1-Nitronaphthalene
2. 3-Nitronaphthalene
3. 3-Nitrobiphenyl
4. 9-Nitroanthracene
5. 2-Nitrofluoranthene
6. 1-Nitropyren
7. 2-Nitropyrene

The authors showed that ambient concentrations of the more volatile PAHs and nitroarenes can be far greater than those of the less volatile species, suggesting that the most abundant nitroarenes in ambient air are from atmospheric transformations of PAHs emitted from combustion sources.

PAHs in Air and Soil

The pattern of nine polycyclic aromatic hydrocarbons from air samples and the pattern from soil samples were compared by Vogt et al.[7] Air samples were collected for 24 hours by a high-volume sampler and analyzed using high-resolution gas chromatography (HRGC). Soil samples, collected from 12 locations in Norway, were analyzed for the nine PAHs.

The nine PAHs were

1. Naphthalene
2. Acenaphthene
3. Biphenyl
4. Fluorene
5. Phenanthrene

6. Fluoranthene
7. Pyrene
8. Chrysene/triphenylene
9. Benzo[a]pyrene

Thrane et al.[8] described the sampling and analysis of the samples. The study was intended to investigate the possibility of sources of input of PAHs to the soil of Norway. Comparison of the pattern of unsubstituted PAHs in air and in soil showed that there were clear differences in these patterns.

The authors[7] concluded that "analyzing soil samples that come from areas with a dominating local input source and comparing the pattern of PAHs in these samples to that of air samples from the same area may allow interpretation of the mechanisms that govern the distribution of PAHs between soil and air."

Formation and Occurrence of Methylnitronaphthalenes in Ambient Air

The nitro isomers formed from the gas-phase reactions of the polycyclic aromatic hydrocarbons 1- and 2-methylnaphthalene with the OH radical (in the presence of NO_x) and with N_2O_5 have been investigated.[9] The methylnitronaphthalenes measured from these gas-phase reactions were compared with methylnitronaphthalenes observed in air samples.

The gas-phase reactions were made in a 6400-L all-Teflon environmental chamber in black-light irradiation. The PAHs flowed into the chamber in nitrogen. Prior to and after the reactions, gas chromatography with flame ionization detection (GC-FID) was used to monitor the gas-phase PAHs concentrations.

A Tenax GC cartridge was used to collect PAHs from a 100-mL sample of the chamber volume. The adsorbed gas was thermally desorbed at 250°C from the cartridge onto the head of a 15-m DB-5 megabore column held at 60°C. The column was temperature programmed at 8°C/min. The FID peak areas were used for quantification.

Five replicate experiments were performed for each isomer. Prior to each experiment, the environmental chamber was cleaned by flushing for several hours with dry air while maintaining the black lights at maximum intensity.

The methylnitronaphthalene isomers and other products were identified using a gas chromatograph interfaced with a mass selective detector (MSD). The following three different silica capillary columns were used: (1) a 25-m SB-Smetic column, (2) a 30-m DB-17 column, and (3) a 50-m HP-5 column.

Ambient air samples were collected in Glendora, CA (about 20 km northeast of downtown Los Angeles) on 9 days. They were collected by modified high-volume samplers with four polyurethane foam plugs beneath a filter. The methylnitronaphthalenes were analyzed by GC-MS-MID (multiple ion detection) using the MSD.

The laboratory data and the ambient measurements showed the importance of atmospheric gas-phase PAHs for the formation of ambient nitroarenes. The authors concluded that "atmospheric reactions of PAHs must be taken into account for risk assessments and for control strategy decisions."

PAHs in the Silicon Carbide Industry

Dufresne et al.[10] described methods for identifying and quantifying the main constituents of dust in the silicon carbide industry and for assessing workers' exposure in two silicon carbide plants.

X-ray diffraction analysis was used to determine the airborne dust content of various polymorphs of silica, especially quartz, cristobalite, and tridymite. Graphite was excluded by modification of the analytical method.

A sampling device that combined a glass fiber filter and a solid Chromosorb 102 adsorbent tube was used to trap both the gaseous polycyclic hydrocarbons and particulates. A gas chromatograph-mass spectrometer (GC-MS) was used in the identification of the PAHs. The following PAHs were identified

1. Naphthalene
2. Biphenyl
3. Acenaphthene
4. Fluorene
5. Phenanthrene
6. Anthracene
7. Fluoranthene
8. Pyrene
9. Chrysene
10. Benzo[a]anthracene
11. Benzo[e]pyrene
12. Benzo[a]pyrene

Polycyclic aromatic hydrocarbons were measured at each work station in silicon carbide plants. PAHs were produced during the silicon carbide synthesis, but were strongly adsorbed by graphite. The environmental results were to be used to try to correlate observed pulmonary diseases to known contaminants.

PAHs and Their Derivatives in an Eight-Home Study

Wilson et al.[11] developed a quiet, transportable, relatively unobtrusive indoor sampler that was capable of operating at a flow rate (8 ft³/min) that was comparable to the PS-1 outdoor sampler. A small field experiment was conducted in eight homes in Columbus, OH during a winter heating season to evaluate the performance of the sampler.[12] Other objectives of the study were

1. To characterize the indoor concentration profiles of PAHs and PAH derivatives
2. To investigate the influence of several indoor contamination sources on the indoor pollution levels
3. To evaluate the correlations of the concentrations of phenanthrene, pyrene, and fluoranthene with the concentrations of less abundant target compounds

The paper discussed the field performance of the quiet sampler and the analytical results of the pilot field study.

The following compounds were measured:

1. Naphthalene
2. Acenapththylene
3. Phenanthrene
4. Anthracene
5. Fluoranthene
6. Pyrene
7. Cyclopenta[c,d]pyrene
8. Benzo[a]anthracene
9. Chrysene
10. Benzofluoranthenes
11. Benzo[e]pyrene
12. Benzo[a]pyrene
13. Indeno[1,2,3-c,d]pyrene
14. Benzo[g,h,i]perylene
15. Coronene
16. Nicotine
17. Quinoline
18. Isoquinoline
19. Fluorenone
20. Naphthalene-1,8-dicarboxylic acid anhydride
21. 9-Nitroanthracene isomer
22. 9-Nitroanthracene
23. 9-Nitrophenanthrene
24. Pyrene carboxaldehyde
25. 2-Nitrofluoranthene
26. 1-Nitropyrene
27. Benz[a]anthracene-7,12-dione
28. Pyrene-3,4-dicarboxylic acid anhydride

Two of the eight homes were sampled twice to estimate the day-to-day variability of the levels of PAHs and their derivatives within a given home. The selection of the homes gave a large range of the number of cigarettes smoked.

The air sampling module consisted of a quartz fiber filter in series with XAD-4. Both indoor and outdoor samplers were calibrated to 8 ft³/min with the sampling module in place prior to sampling. The gas chromatography-electron capture detection (GC-ECD) time profile of injected sulfur hexafluoride was used to determine the air exchange rate of each house.

The details of preparing sample extracts are given in the paper. Analysis of the sample extracts was by gas chromatography-mass spectrometry (GC-MS) with positive chemical ionization (PCI) to determine PAHs, quinoline, isoquinoline, and nicotine and by GC-MS with negative chemical ionization (NCI) to determine NO_2-PAHs and OXY-PAHs.

The gas chromatography parameters were

1. Column — 50 m × 0.31 mm fused silica capillary column
2. Column coating film thickness — 0.51 μm
3. Temperature program — 45°C for 2 min, raised to 100°C in 5 min, then to 320°C at 6°C/min

Five field blanks were collected to assess the quality of the data and 30 additional samples were collected at 8 homes. The results showed that the quiet sampler could be used to reliably collect samples for an 8-hour duration at a very stable flow rate of 8.0 ± 0.1 ft³/min.

Naphthalene was the most abundant of the PAHs found in indoor and outdoor air; the least abundant was cyclopenta[c,d]pyrene. Highest indoor concentrations for all PAHs were from a home during the sampling period during which the highest number of cigarettes, a total of 20, was smoked. To measure the expected very high levels of nicotine detected in indoor environments in the presence of environmental tobacco smoke (ETS), it was necessary to dilute some sample abstracts by a factor of 10 to 200 and reanalyze them by PCI GC-MS. Except for naphthalene dicarboxylic acid anhydride, pyrene dicarboxylic acid anhydride, and 2-nitrofluoranthene, average indoor levels were higher than average outdoor levels.

Environmental tobacco smoke has the most significant influence on indoor pollution levels. For three types of homes occupied by nonsmokers, those having gas heating and cooking appliances had the highest average concentrations of most PAH compounds.

The concentrations of other target compounds correlated well with the concentrations of the PAHs marker compounds, phenanthrene, fluoranthene, and pyrene. The authors concluded that quinoline and isoquinoline can be used as markers for nicotine for indoor measurement of environmental tobacco smoke.

REFERENCES

1. Thrane, K. E., and A. Mikalsen. "High-Volume Sampling of Airborne Polycyclic Aromatic Hydrocarbons Using Glass Fiber Filters and Polyurethane Foam," *Atmos. Environ.* 15:909 (1981).
2. Lesage, J., G. Perrault, and P. Durand. "Evaluation of Worker Exposure to Polycyclic Aromatic Hydrocarbons," *Am. Ind. Hyg. Assoc. J.* 48:753 (1987).
3. Suess, M. J. "The Environmental Load and Cycle of Polycyclic Aromatic Hydrocarbons," *Sci. Total Environ.* 6:239–250 (1976).

4. National Institute of Occupational Safety and Health. "Benzene-Soluble Compounds in Air," *Manual of Analytical Methods,* 2nd ed. Vol. 1 (P&CAM 217). (Cincinnati, OH: 1976).

5. Alcan International Limited. "Method of Analysis: Method 1190–84," SECAL, Case Postale 370, Arvida, Quebec, Canada.

6. Arey, J., B. Zielinska, R. Atkinson, and A. M. Winer. "Polycyclic Aromatic Hydrocarbon and Nitroarene Concentrations in Ambient Air During a Wintertime High-NO$_x$ Episode in the Los Angeles Basin," *Atmos. Environ.* 21:1437 (1987).

7. Vogt, N. B., F. Brakstad, K. Thrane, S. Nordenson, J. Krane, E. Aamot, K. Kolset, K. Esbensen, and E. Steinnes. "Polycyclic Aromatic Hydrocarbons in Soil and Air: Statistical Analysis and Classification by the SIMCA Method," *Environ. Sci. Technol.* 21:35 (1987).

8. Thrane, K., A. Mikalsen, and H. Stray. *Int. J. Environ. Anal. Chem.* 23:111 (1985).

9. Zielinska, B., J. Arey, R. Atkinson, and P. A. McElroy. "Formation of Methylnitronaphthalenes from the Gas-Phase Reactions of 1- and 2-Methylnaphthalene with OH Radicals and N$_2$O$_5$ and Their Occurrence in Ambient Air," *Environ. Sci. Technol.* 23:723 (1989).

10. Dufresne, A., J. Lesage, and G. Perrault. "Evaluation of Occupational Exposure to Mixed Dusts and Polycyclic Aromatic Hydrocarbons in Silicon Carbide Plants," *Am. Ind. Hyg. Assoc. J.* 48:160 (1987).

11. Wilson, N. K., M. R. Kuhlman, J. C. Chuang, G. A. Mack, and J. E. Howes, Jr. "A Quiet Sampler for the Collection of Semi-Volatile Organic Pollutants in Indoor Air," *Environ. Sci. Technol.* 23:1112 (1989).

12. Chuang, J. C., G. A. Mack, M. R. Kuhlman, and N. K. Wilson. "Polycyclic Aromatic Hydrocarbons and Their Derivatives in Indoor and Outdoor Air in an Eight-Home Study," *Atmos. Environ.* 25B:369–380 (1991).

CHAPTER 10

Formaldehyde

INTRODUCTION

Uses of Formaldehyde

Formaldehyde is an important industrial chemical. It is widely used[1] for the manufacture and chemical treatment of a variety of products such as cotton, explosives, fertilizers, resins, rubber, and in the synthesis of many organic compounds. It also has been used in the cosmetic and textile industries.[2,3]

Need for Collection and Analysis

The toxicity of formaldehyde has been studied[3,4] and it has been investigated as a suspected carcinogen.[4] It is one of the organic pollutants most often found in air. The development of new methods for monitoring occupational and environmental exposure to formaldehyde is of great interest from the analytical and toxicological viewpoints. The need to quantify efficiently and reliably the formaldehyde in air exists at formaldehyde levels well below 1 ppm. Adverse health effects have been attributed to formaldehyde exposure in occupational and residential environments, but a causal relationship between formaldehyde presence and adverse health effects had not been rigorously established.[5]

Exposure

Occupational exposure to formaldehyde occurs in the range 0.1 to 5 mg/m^3 of formaldehyde in air,[6] and levels of 0.01 to 0.1 mg/m^3 are often found in offices and homes.[7]

COLLECTION AND MEASUREMENT OF FORMALDEHYDE IN AIR

Numerous methods have been used to collect and measure (quantify) formaldehyde in air.[8] Formaldehyde has been measured directly, and concentration techniques using liquid or solid adsorbents have been used. Detection procedures have included polarography, liquid chromatography, ion chromatography, gas-liquid chromatography, length of stain, colorimetry, and others.[8]

The most common methods for sampling formaldehyde in workplace and residential environments[8] include: (1) the National Institute for Occupational Safety and Health (NIOSH) impinger/chromotropic acid method,[9] (2) the modified impinger/pararosaniline method,[10] and (3) length-of-stain detector tubes.[11]

The detector tubes had not been used widely in residential sampling because the detection limits for the tubes exceeded typical residential levels in most cases, and the sensitivity of the method was not that desired. Mean indoor formaldehyde concentrations in air measured in two surveys of "typical" Houston, TX houses,[7,12] and in a study of Texas mobile homes,[13] were 0.07, 0.08, and 0.15 ppm, respectively. All of these three levels were below the detection limit (0.2 ppm) of the most widely used detector tube for formaldehyde measurement — the Draeger 0.2/a detector tube.

Draeger had extended the range of measurement to 0.04 to 0.05 ppm formaldehyde in air by using a "precleansing" activation tube with the 0.2/a indicating tube and increasing the total volume sampled by a factor of 5.[11] Because of the simplicity and the relatively low cost of the Draeger tube, Beck and Stock[8] considered that it might be an attractive alternative to other methods if the performance of the tube was found to be acceptable.

Tests of the Performance of Four Different Formaldehyde Monitoring Devices

Laboratory and field tests of the performance of four different formaldehyde monitoring devices were described by Coyne et al.[14] The devices evaluated were

1. A formaldehyde badge
2. An impregnated charcoal tube
3. An impinger/polarographic system
4. A formaldemeter

The formaldehyde badge combined a patented multicavity diffuser element with a specific absorbing sodium bisulfite solution in a self-contained package. The impregnated charcoal tube contained 150 mg of charcoal impregnated with a proprietary oxidating agent, in two sections. The impinger/polarographic system had two standard glass midget impingers, each containing 15 mL of extra-purified distilled deionized water. Polarographic reduction of the hydrazine derivative was used to analyze the samples collected in the impingers. An electrochemical fuel cell containing two platinum electrodes to detect and measure formaldehyde concentrations was used in the formaldemeter. The monitoring devices were exposed to known formaldehyde concentrations.

The formaldehyde badge had a sensitivity of 2.8 ppm-hour and accurately determined time-weighted average (TWA) exposures; however, the authors reported that it was not sufficiently sensitive to measure short-term exposure limit (STEL) exposures, whereas, positive interferences resulted if 1-3-butadiene was present.

The impregnated charcoal tube method had a sensitivity of 0.06 ppm, based on a 25-L air sample, and accurately determined the STEL.

The impinger/polarographic method had a sensitivity of 0.06 ppm, based on a 20-L air sample, and accurately determined the STEL. However, it was found to be not very practical for TWA personal monitoring measurements.

The formaldemeter had a sensitivity of 0.2 ppm. It, however, responded to many interferences.

Study of Three Methods for Measuring Low-Level Formaldehyde Concentrations

The chromotropic acid method, the modified pararosaniline method, and the Draeger detector tube method for measuring low-level formaldehyde concentrations (less than 1 ppm) typically found in indoor air were compared by Beck and Stock.[8]

Samples were collected from particleboard, formalin, and conventional housing sources. Relatively high, medium, and low concentrations of formaldehyde were generated and sampled for the formalin and particleboard sources. Medium and low levels of formaldehyde were sampled for the conventional house. Levels of 0.4 to 0.8 ppm, 0.1 to 0.3 ppm, and less than 0.1 ppm were considered to be high, medium, and low formaldehyde levels, respectively. Four replicate samples for each of the three methods were collected for each source level.

Several days prior to sampling, a 0.6 m × 0.6 m × 0.6 m (8 ft³) box constructed of particleboard was placed in the laboratory to reach temperature and relative humidity equilibrium. Using formalin as a source, test atmospheres were prepared by adding 37% (weight/weight) reagent grade formalin to 50-L Tedlar bags. Air was metered into the bags to approximate the desired concentration. The test bag was prepared 1 day prior to sampling.

Colorimetric analyses were made using a grating UV/VIS spectrophotometer with matched quartz cuvettes. The reference solution for all absorbance measurements was reagent water prepared by filtration through an activated carbon column followed by a mixed-ion exchange column.

A modified NIOSH Method 3500[9] was used for the chromotropic acid method, in which formaldehyde is collected in a series of two midget impingers containing a 1% sodium bisulfite solution. Subsequent color development used chromotropic acid and sulfuric acid. The modifications to the NIOSH method were

1. Paraformaldehyde was used to prepare formaldehyde stock solution, the solution was standardized using the sodium bisulfite titration method.[10]
2. Each day of the analysis, a fresh 5% chromotropic acid solution was prepared.
3. After addition of chromotropic acid solution, samples stood for 5 min prior to addition of sulfuric acid.
4. To scrub sulfur dioxide generated from the first two impingers, a third impinger in series contained 0.3 N hydrogen peroxide.
5. Since heat generated from the addition of the sulfuric acid was sufficient for color development, the heating step at 95°C was omitted.

The modified pararosaniline method was used.[7] In this method, formaldehyde is collected in two impingers in series containing reagent water. Pararosaniline/hydrochloric acid and sodium sulfite reagents were used for subsequent color development[15] Standards, which were treated in the same manner and at the same time as the samples, were prepared for each set of samples analyzed. In the modified pararosaniline method, sulfur dioxide was a negative interferent.

The reaction mode of the Draeger tube is the reaction of formaldehyde, xylene vapor, and sulfuric acid to form a pink compound. Short-term formaldehyde (0.2/a) tubes and a Draeger bellows hand pump were used for the Draeger procedure. A precleansing activation tube was used upstream of the indicating tube for concentrations below 0.5 ppm formaldehyde. Samples were collected in air sufficiently low in water vapor content that the acceptable range of water vapor content for the tubes was not exceeded.

Relative humidity during sample collection was measured using a sling psychrometer.

For each of the three methods of analysis, four samples were collected. All samples were analyzed within 24 hours of collection, many on the day of collection. On collection days, a series of standard concentrations were prepared and stored in the same manner as the samples. For high-level formaldehyde concentrations, only the 0.2/a Draeger tube was required for sampling. An activation tube was used in conjunction with the 0.2/a tube for medium- and low-level concentrations.

The collection efficiency, defined as the quantity of formaldehyde collected in the first impinger divided by the total quantity, expressed as a percentage, per impinger for the pararosaniline method was 87.4% mean. The second impinger, for every case for the chromotropic acid method, was below the minimum detectable concentration. A total of 12 calibration curves (absorbance vs. formaldehyde concentration) was generated for each colorimetric method. The pararosaniline method was indicated to be more sensitive than the chromotropic acid method since the slope for the pararosaniline calibration curve was more than double that for the chromotropic acid method. For both procedures, there was a larger relative uncertainty at the low end of the calibration curve.

The following interpretations of results were made:[8]

- "The Draeger tube method using an activation tube gives lower results than either of the impinger methods. Without using an activation tube (concentrations >0.5 ppm), the Draeger tube method was comparable to the two impinger methods."
- "There are indications that the chromotropic acid method gives different results than the modified pararosaniline method, depending on the source of formaldehyde."
- "The modified pararosaniline method indicated higher results than the chromotropic acid method when sampling from a particleboard box but not from a formalin source."
- "Overall analytical precision for each method of analysis was good."

Evaluation of a Passive Diffusive Sampler

An extensive evaluation of a passive diffusive sampler utilizing a glass fiber filter impregnated with 2,4-dinitrophenylhydrazine (DNPH) and phosphoric acid and mounted into a modified aerosol-sampling cassette was reported by Levin et al.[16]

In the application of the diffusive sampler, formaldehyde hydrazone formed was desorbed and determined by high-performance liquid chromatography and ultraviolet detection. The sampling rate and the effects of formaldehyde concentration, sampling time, air velocity, and air humidity on the sampling rate were determined. The sampler was also evaluated in the field.

For diffusive sampling, the coated glass fiber filters were prepared by a procedure[17] modified to include glycerine in the coating section. The filters were used in two-section polypropylene filter holders. The freshly coated filters were dried for 30 min at room temperature and conditioned as were filters for active sampling. The length of the controlled diffusion path was 10 mm and the front cross-sectional area of the sampler was 7.1 cm.[2]

In a dynamic system, formaldehyde was generated by decomposition of paraformaldehyde at 25 to 30°C. In the dissection room of a hospital, personal sampling tubes were used for active sampling. With the diffusive samplers, parallel samples were taken.

The high-performance liquid chromatography procedure was the following:

1. Formaldehyde 2,4-dinitrophenylhydrazone was eluted from both 13-mm and 37-mm filters by shaking with 3 mL of acetonitrile in a 4-mL glass vial for 1 min
2. After the solution had been filtered, 10 μL was injected into the liquid chromatograph
3. Details of the formaldehyde determination had been described previously[17]
4. Hydrazone was detected at 365 nm, with a detection limit of approximately 0.5 ng corresponding to 0.07 ng of formaldehyde

Recovery of gas-phase-generated formaldehyde as 2,4-dinitrophenylhydrazone from 13-mm DNPH-impregnated filters ranged between 81 and 103%. In all experiments with the passive sampler, the active sampling method was used as the reference method. The coated 13-mm filters made it possible to determine formaldehyde down to approximately 1 μg/m³ in a 50-L air sample. The diffusive sampler was exposed to formaldehyde concentrations of 0.1 to 5.0 mg/m³. The sampling rate was not affected by the relative humidity of the sampler air, the formaldehyde concentration within the range studied, or DNPH load between 5 and 8.5 mg. The reproducibility of the diffusive sampling method was better than 3%. The sensitivity of a 15-min sample, with a formaldehyde blank of 0.15 μg per filter, was approximately 0.2 mg/m³; for an 8-hour sample, the sensitivity would become 0.005 mg/m³ (5 ppb).

The diffusive sampler was evaluated in a field investigation in a dissection room of a hospital where medical students were dissecting human organs embalmed in formalin. The sampling time was 15 to 150 min and the formaldehyde level ranged from 0.1 to 0.5 mg/m³. The results obtained with passive and active sampling were in good agreement. The diffusive sampler was found to be very useful for monitoring formaldehyde exposure in working environments and in homes.

A Diffusion Colorimetric Air Monitoring Badge

Kring et al.[18] presented additional laboratory validation work on a diffusion colorimetric air monitoring badge [19] designed for a broad sampling range. They also presented results for field verification tests and some observations on the most widely used (then) independent method—NIOSH P&CAM 125.

Four independent sampling methods were used in all laboratory studies:

1. The pump/impinger sampling method, NIOSH P&CAM 125
2. The pump/silica gel tube monitoring method
3. The CEA Model 555 monitor
4. The Lion formaldemeter

The field test protocol was based on NIOSH sampling strategy, a verification plan developed jointly with the Ethylene Oxide Industry Council and past experience.

Laboratory validation tests confirmed an overall accuracy of ± 9.6 to 11.5% for the badge over the exposure range of 0.12 to 6.8 ppm. Laboratory test data also showed that the badge met NIOSH and OSHA accuracy requirements of 25% or less down to 1.6 ppm-hour (200 ppb) for an 8-hour TWA exposure, was capable of accurately sampling for 15-min exposures, and was not affected by pressure or relative humidity variations.

Results for field tests at three different plant sites showed the badge to have excellent correlation with three commonly used methods for monitoring formaldehyde.

Study of the Accuracy of the Draeger Tube

A brief study to determine the accuracy of the Draeger tube when the analyst was the only variable was conducted by Balmat.[20] In this study, the atmosphere monitored was maintained at known and constant formaldehyde concentration, temperature, and relative humidity.

Formaldehyde concentrations of 0.5 to 3 ppm were generated by a formaldehyde generator[21] which produced constant formaldehyde levels in a flowing air stream. A modification of the chromotropic acid method[22] was used to determine the concentration of formaldehyde in air; the accuracy of reported values was ±3% relative. The monitored air was at 24°C and 11% relative humidity during all runs.

Eight analysts who were research chemists and technicians knowledgeable in analytical techniques and methodology monitored the formaldehyde levels. They were instructed in the use of the Draeger tubes in accordance with the manufacturer's instructions. The usable range of the Draeger tubes was claimed to be 0.5 to 10 ppm.

The analysts monitored three levels of formaldehyde concentration: 2.80, 1.00, and 0.5 ppm. The range of relative standard deviation was 15 to 60%. The absolute mean bias (observed mean formaldehyde concentration − known formaldehyde concentration)/(known formaldehyde concentration) varied from +0.89 to +1.6, i.e., 89 to 160% higher than the known formaldehyde concentration. Out of the 19 individual values, 2 were equal to or lower than the known values.

Balmat concluded that:[20]

"It is recognized that the amount of data collected during this study is not sufficient for a rigorous statistical evaluation: however, the study does indicate that monitoring for formaldehyde with the Draeger tube will produce excessively high results."

Determination of Formaldehyde in Air by Two Flow Injection Methods

The determination of formaldehyde by two flow injection (FI) methods, conventional and stopped-flow, were described by Munoz et al.[1] The methods are based on the reaction of formaldehyde with pararosaniline in the presence of sulfite ion. Spectrophotometric detection of formaldehyde was made at 570 nm. Formaldehyde in air in work environments was determined by the stopped-flow mode.

The experimental apparatus consisted of:

1. A flow injection analyzer equipped with a variable-volume injection valve and two four-channel peristaltic pumps
2. A microprocessor
3. A spectrophotometric detector
4. A recorder
5. A glass flow cell with an inner volume of 18 μL and a pathlength of 10 mm

The flow injection system was thermostatted at $30 \pm 3°C$.

The system was used to analyze air from a research laboratory handling formaldehyde. The air was drawn through 50 mL of water at 1 L/min for 40 min to collect formaldehyde. This solution was transferred to a 100-mL calibrated flask and made up to the mark with water.

Then, 80 μL of the sample solution was injected directly into a pararosaniline-hydrochloric acid carrier which merged with a sodium sulfite solution. The peak height for the conventional flow injection method or the signal increment for the stopped-flow method was measured at 570 nm. The stopped-flow method was applied to the aqueous sample described above to determine formaldehyde in air. Calibration graphs for aqueous formaldehyde standard solutions were obtained.

The detection limit in the stopped-flow method was 0.1 μg/mL of formaldehyde. In both the conventional and stopped-flow methods, sampling rates of 41 and 18 samples per hour, respectively, were achieved. The results for the standard chromotropic acid method were used to confirm the results for the stopped-flow method applied to the determination of formaldehyde in air in work environments. The flow injection methods were simple and rapid.

Methods for Monitoring Nanogram Levels of Formaldehyde

A method for monitoring exposure to nanogram levels of formaldehyde, and tests thereof, were described by Bisgaard et al.[23] The method was the basis of a commercial sampling tube. It involved collection on Chromosorb W coated with 0.6% 7-amino-5-hydroxy-2-naphthalenesulfonic acid in concentrated sulfuric acid, desorption with concentrated sulfuric acid, and colorimetric or fluorimetric determination of the reaction product.

The standard sampling tube consisted of three sections:

1. The first section, which contained a drying agent to remove water vapor from the air sample
2. The second section, which contained a solid sorbent material which formed a fluorescent compound when exposed to formaldehyde
3. A backup section which contained the same sorbent material

The sampling tubes were exposed to formaldehyde concentrations in the range 0.2 to 0.8 mg/m^3.

Most tests for the evaluation of the sampling and analytical principle were performed in dry air and without drying sections, leaving the problem of finding a suitable desiccant to be solved later. Studies were made of the capacity, interference, and stability of the formaldehyde samples.

The authors concluded that both colorimetric and fluorimetric determinations in acid solution were adequate for small amounts of formaldehyde in air. However, if measurements of more than 15-min duration were to be taken, the adequacy of the desiccant still remained to be verified.

Coated Silica Gel Cartridge Method for Determination of Formaldehyde in Air: Identification of an Ozone Interference

Substantial negative interference due to the determination of formaldehyde in air by the 2,4-dinitrophenylhydrazine (2,4-DNPH-)-coated silica gel cartridge technique of Tejada[24] were identified and quantified by Arnts and Tejada.[25] A systematic study was made of the accuracy of the 2,4-DNPH-silica gel cartridge (SGC) method for the measurement of formaldehyde in the presence of ozone.

A dynamic dilution apparatus simultaneously delivered 50% ambient air and 50% cleaned ambient (zero) air blended with formaldehyde, or 100% zero air blended with formaldehyde, to two heated glass sampling manifolds. Prior to introduction of formaldehyde, the zero air delivered to one manifold passed through a mercury lamp ozone generator. In this manifold, controlled amounts of ozone could be introduced without photolysis of aldehydes or NO$_2$. The mixtures reflected ambient air conditions since moisture was not removed from the ambient air.

Two cartridges and two impinger samples were collected for 120 min simultaneously at each manifold. Analyses of the cartridges and impinger samples were made by a shortened version of the high-pressure liquid chromatography (HPLC) procedure of Tejada.[24] An acetonitrile, water, and methanol program was used to check for coeluting peaks. Three sets of experiments were carried out at nominal formaldehyde concentrations of 20, 40, and 140 ppb. At each of these formaldehyde concentrations, sampling was performed with increments of ozone of 0, 120, 300, 500, and 770 ppb above background. Zero air plus formaldehyde and 100% air experiments were performed to ascertain the source of unknown HPLC peaks.

The losses of ability to measure formaldehyde concentrations were quite severe for the coated silica gel cartridges. At an increment of 120 ppb of ozone and about 25 ppb of formaldehyde, between 44 and 52% of the formaldehyde was not measured; at 300 ppb of ozone (often seen in severe urban smog episodes), between 57 and 61% of the formaldehyde was not measured. The impinger data, however, showed no formaldehyde-2,4-DNPH loss as a function of ozone over the same experimental conditions.

A few experiments were performed using the 2,4-DNPH cartridge technique of Kuwata et al.,[26] who used a C_{18} Sep-Pak cartridge which contained silica gel which had been passivated with a nonpolar organic phase.

Ozone reacted with 2,4-DNPH and its formaldehyde hydrazone on the surface of silica gel but not on the nonpolar C_{18} coated silica gel surface. The reactivity of the adsorbed molecules toward ozone was apparently influenced by the surface of the solid substrate. Arnts and Tejada[25] recommended the suspension of the use of 2,4-DNPH-silica gel for ambient monitoring unless a carbonyl-passive ozone scrubber was employed.

Solid Sorbent Tube Collection and Ion Chromatographic Analysis

A method for the collection and analysis of atmospheric formaldehyde was discussed by Kim et al.[27] The method was based on solid sorbent tube collection and ion chromatographic analysis. Ion chromatography consisted of ion-exchange chromatography, background ion suppression, and conductimetric detection.

The solid sorbent tube was prepared with two sections of impregnated charcoal separated by silylated glass wool. The proprietary impregnating material converted formaldehyde to formate.

The collection and analysis procedures consisted of the following:

1. After sampling, at either 50 cm³/min or 200 cm³/min, each section of the charcoal was placed in a separate centrifuge tube
2. Then, 10 mL of 0.1% hydrogen peroxide solution was added to the adsorbent section and 5.0 mL was added to the backup section
3. After 1 hour of occasional shaking, each sample was placed in an ultrasonic bath for 20 min
4. After filtration through a mixed cellulose ester filter, a fraction of filtrate was injected into the anion system of an ion chromatograph

The overall recovery of laboratory-generated samples was 100%, with 11% relative standard deviation.

The DTNB Method

Hoogenboom et al.[5] described a novel method for the determination of formaldehyde in air — an alternative to other available colorimetric methods. The method was based on the determination of sulfur dioxide in air[28] as aqueous sulfite or bisulfite by the use of 5,5′-dithio-bis(2-nitrobenzoic acid) (DTNB).[29] It was claimed to be more sensitive than the chromotropic acid method and less sensitive than the modified pararosaniline method. Interferences were claimed to be no more serious than those for the chromotropic method, the pararosaniline method, or the 3-methyl-2-benzothiazolone hydrazone method.[30]

Standards for the determination of the calibration line, mean absorbance vs. formaldehyde concentration, were prepared. A UV-VIS spectrometer with a slit width of 1.0 nm was used to measure absorbance.

Formaldehyde in air samples was determined by the DTNB method. Known concentrations of gaseous formaldehyde in air was absorbed in pH 7 buffer prior to the determinations. At the 95% confidence level, the precision and accuracy of the method were indicated by the mean percentage recovery (99.9 ± 2.7%), the mean percentage standard error (1.72 ± 1.0%), the mean percentage relative error (5.04 ± 3.2%), and the mean percentage absolute error (1.86 ± 1.6%). The slope of the graphic representation of formaldehyde air concentration found vs. formaldehyde air concentration was 0.993.

Formaldehyde in Urban Pollution in Brazil

Grosjean et al.[31] measured ambient levels of carbonyls in three urban areas of Brazil: Sao Paulo, Rio de Janeiro, and Salvador.

To be used for sampling for carbonyls, small Sep-Pak C_{18} resin cartridges were cleaned and wetted with HPLC-grade methanol and acetonitrile, were impregnated with an acidic solution of twice recrystallized 2,4-dinitrophenylhydrazine (DNPH) in HPLC-grade acetonitrile, and were then allowed to dry under vacuum in a desiccator.

At flow rates of 0.5 to 2.0 L/min, air was sampled by connecting the DNPH-impregnated cartridge to a calibrated flowmeter and an air sampling pump. Airborne carbonyls reacted with DNPH to form the corresponding 2,4-dinitrophenylhydrazones, which were retained on the cartridges. Subsequently in the laboratory, the cartridges were eluted slowly with 2 mL of HPLC-grade acetonitrile and the acetonitrile extracts were analyzed directly by liquid chromatography with ultraviolet detection.

Twelve carbonyls and groups of unresolved isomers were identified in air from the three urban areas:

1. Acetaldehyde (the most abundant)
2. Formaldehyde (the second most abundant)
3. Acrolein
4. Acetone
5. Propanal
6. Unsaturated C_4 aliphatics
7. n-Butanal + 2-butanone
8. Benzaldehyde
9. C_5 aliphatics
10. Glyoxal
11. Tolualdehyde isomers
12. Methylglyoxal

The overall pattern of the chromatograms was similar for all three urban areas, suggesting a common source — vehicle emissions, for example. Indoor concentrations of carbonyls in public buildings were similar to outdoor levels.

Ambient concentrations of formaldehyde were in the range of 1 to 34 ppb. The concentrations of acetaldehyde were in the range of 1 to 35 ppb. Ambient levels of acetone were up to 20 ppb, and levels of other carbonyls were all less than 5 ppb.

Comparison of Chromotropic Acid and Modified Pararosaniline Methods

The continuous, modified pararosaniline method and the chromotropic acid method for determining formaldehyde concentrations in measuring chambers containing particleboard under controlled temperature and relative humidity were compared by Silberstein.[32]

The methods used to calibrate the analyses were also compared. The modified pararosaniline method was calibrated using an airborne formaldehyde standard generated by permeation tubes .[33] The chemotropic acid method was calibrated using a formalin standard.[34]

Dynamic measuring chambers were used in Silberstein's study to determine formaldehyde concentrations of emissions from individual pressed-wood products. The measuring chambers were located in an environmental chamber maintained at 23°C and 50% relative humidity. In a control experiment, formaldehyde was injected into each chamber at a known rate using gas standard generators containing polyoxymethylene permeation tubes heated to 100°C. At the outlet of each chamber, formaldehyde concentrations were measured by the pararosaniline and chromotropic acid methods. By varying the air exchange rate, a range of formaldehyde concentrations was obtained.

The continuous, modified pararosaniline analysis was made using an air monitor fitted with a formaldehyde analytical module. Air was drawn continuously through the monitor at a fixed flow rate between 0.5 and 1 L/min. The air was scrubbed with 0.02% pararosaniline in 0.1 N hydrochloric acid, 0.5% sodium sulfite was added at a volume ratio of 1:2, and color was allowed to develop for about 10 min and the absorbance was monitored at 570 nm.

NIOSH Method 3500[34] was used for the collection of samples by impinger and analysis by the chromotropic acid method. Two samples were collected for approximately 1 hour for each chamber. Two analyses were made for each sample. A few double-impinger tests were made using two impingers in series. The chromotropic acid analyses were calibrated using the formalin titration method.

Formaldehyde concentrations determined by the two methods agreed on both synthetic formaldehyde atmospheres and on emissions from five lots of particleboard from four different manufacturers. The results of the study suggested that both methods are suitable for measurements of pressed-wood products under controlled temperature and relative humidity conditions.

REFERENCES

1. Munoz, M. P., F. J. M. de Villena Rueda, and L. M. P. Diez. "Determination of Formaldehyde in Air by Flow Injection Using Pararosaniline and Spectrophotometric Detection," *Analyst* 114:1469–1471 (1989).

2. Gonzalez, E. *Formaldehido: Toxicologia e Impacto Ambiental*, Fundacion Mapfre, Madrid, 1986.

3. Pickard, A. D., and E. R. Clark. *Talanta* 31:763 (1984).

4. "Threshold Limit Values for Chemical Substances and Physical Agents in the Workroom Environment," American Conference of Government Industrial Hygienists, Cincinnati, OH, 1983.

5. Hoogenboom, B. E., R. W. Hynes, C. M. Mann, M. Ekman, C. E. McJilton, and J. B. Stevens. "Validation of a Colorimetric Method for Determination of Atmospheric Formaldehyde," *Am. Ind. Hyg. Assoc. J.* 48:420–424 (1987).

6. Hart, R. W., A. Terturro, and L. Neimith, Eds. *Environ. Health Perspect.* 58:325 (1984).

7. Stock, T. H., and S. R. Mendez. "A Survey of Typical Exposures to Formaldehyde in Houston Area Homes," *Am. Ind. Hyg. Assoc. J.* 46:313–317 (1985).

8. Beck, S. W., and T. H. Stock. "An Evaluation of the Effect of Source and Concentration on Three Methods for the Measurement of Formaldehyde in Indoor Air," *Am. Ind. Hyg. Assoc. J.* 51:14–22 (1990).

9. National Institute for Occupational Safety and Health, *Manual of Analytical Methods,* 3d ed. (NIOSH Publ. No. 84–100). Cincinnati, OH: National Institute for Occupational Safety and Health, 1984, pp. 3500–1 to 3500–4.

10. Miksch, R. A., D. W. Anthon, L. Z. Fanning, C. D. Hollowell, K. Revzan, and J. Glanville. "Modified Pararosaniline Method for the Determination of Formaldehyde in Air," *Anal. Chem.* 53:2118 (1981).

11. Draegenwerk AG Lubeck: *Detector Tube Handbook.* 6th ed. Lubeck, Federal Republic of Germany: Draeger, 1985.

12. Stock, T. H. "Formaldehyde Concentrations Inside Conventional Housing," *J. Air. Poll. Control Assoc.* 37:913–918 (1987).

13. Stock, T. H., R. M. Monsen, D. A. Sterling, and S. W. Norsted. "Indoor Air Quality Inside Manufactured Housing in Texas," Paper 85–85.1, presented at the 78th Annual Meeting of the Air Pollution Control Association, Detroit, MI, June 16–21, 1985.

14. Coyne, L. B., R. E. Cook, J. R. Nann, S. Bouyoucos, O. F. McDonald, and C. L. Baldwin. "Formaldehyde: A Comparative Evaluation of Four Monitoring Methods," *Am. Ind. Hyg. Assoc. J.* 46:609 (1985).

15. Petreas, M., S. Twiss, D. Pon, and M. Imada. "A Laboratory Evaluation of Two Methods for Measuring Low Levels of Formaldehyde in Air," *Am. Ind. Hyg. Assoc. J.* 47:276–280 (1986).

16. Levin, J.-O., R. Lindahl, and K. Andersson. "A Passive Sampler for Formaldehyde in Air Using 2,4-Dinitrophenylhydrazine-Coated Glass Fiber Filters," *Environ. Sci. Technol.* 20:1273–1276 (1986).

17. Levin, J.-O., K. Andersson, R. Lindahl, and C.-A. Nilsson. *Anal. Chem.* 57:1032 (1985).

18. Kring, E. V., G. R. Ansul, A. N. Basilio, Jr., P. D. McGibney, J. S. Stephens, and H. L. O'Dell. "Sampling for Formaldehyde in Workplace and Ambient Air Environments — Additional Laboratory Validation and Field Verification of a Passive Air Monitoring Device Compared with Conventional Sampling Methods," *Am. Ind. Hyg. Assoc. J.* 45:318 (1984).

19. Kring, E. V., D. J. Damrell, A. N. Basilio, Jr., P. D. McGibney, J. J. Douglas, T. T. Henry, and G. R. Ansul. "Laboratory Validation and Field Verification of a New Passive Air Monitoring Badge for Sampling Ethylene Oxide in Air, "*Am. Ind. Hyg. Assoc. J.* 45:697 (1984).

20. Balmat, J. L. "Accuracy of Formaldehyde Analysis Using the Draeger Tube," *Am. Ind. Hyg. Assoc. J.* 47:512–513 (1986).

21. Balmat, J. L. "Generation of Constant Formaldehyde Levels for Inhalation Studies," *Am. Ind. Hyg. Assoc. J.* 46:690–692 (1985).

22. E. I. du Pont de Nemours & Co.: Chemicals and Pigments Dept. Method No. A125.011 (Wilmington, DE).

23. Bisgaard, P., L. Molhave, B. Reitz, and P. Wilhardt. "A Method for Personal Sampling and Analysis of Nanogram Amounts of Formaldehyde in Air," *Am. Ind. Hyg. Assoc. J.* 45:425 (1984).

24. Tejada, S. B. *Int. J. Environ. Anal. Chem.* 26:167 (1986).

25. Arnts, R. R., and S. B. Tejada. "2,4-Dinitrophenylhydrazine-Coated Silica Gel Cartridge Method for Determination of Formaldehyde in Air: Identification of an Ozone Interference," *Environ. Sci. Technol.* 23:1428–1430 (1989).

26. Kuwata, K., M. Uebori, H. Yamasaki, Y. Kuge, and Y. Kiso. *Anal. Chem.* 55:2013 (1983).

27. Kim, W. S., C. L. Geraci, Jr., and R. E. Kupel. "Solid Sorbent Tube Sampling and Ion Chromatographic Analysis of Formaldehyde ," *Am. Ind. Hyg. Assoc. J.* 41:334 (1980).

28. Hoogenboom, B. E., R. W. Hynes, C. E. McJilton, and J. B. Stevens. "Validation of a Simpler Method for Determination of Atmospheric Sulfur Dioxide," *Am. Ind. Hyg. Assoc. J.* 47:552 (1986).

29. Humphrey, R. E., M. H. Ward, and W. Hinze. "Spectrophotometric Determination of Sulfite with 4,4'-Dithiopyridine and 5,5'-Dithiobis(2-Nitrobenzoic Acid)," *Anal. Chem.* 42:69 (1970).

30. Miksch, R. R., and D. W. Anthon. "A Recommendation for Combining the Standard Analytical Methods for the Determination of Formaldehyde and Total Aldehydes in Air," *Am. Ind. Hyg. Assoc. J.* 43:362 (1982).

31. Grosjean, D., A. H. Miguel, and T. M. Tavares. "Urban Air Pollution in Brazil: Acetaldehyde and Other Carbonyls," *Atmos. Environ.* 24B: 101–106 (1990).

32. Silberstein, S. "Comparison of the Chromotropic Acid and Pararosaniline Methods for Measuring Formaldehyde Concentrations of Pressed-Wood Product Emissions," *Am. Ind. Hyg. Assoc.* 51:102–106 (1990).

33. Scaringelli, F. P., A. E. O'Keefe, E. Rosenberg, and J. P. Bell. "Preparation of Known Concentrations of Gases and Vapors with Permeation Devices Calibrated Gravimetrically," *Anal. Chem.* 42:871 (1970).

34. National Institute for Occupational Safety and Health, *Manual of Analytical Methods,* 3rd ed. Ed. P. M. Heller, (NIOSH Publication No. 84–100), Cincinnati, OH, National Institute for Occupational Safety and Health, 1984. Method 3500.

INDEX

Date Due
